NATURAL DISASTERS

EARTH SCIENCE READINGS

Nancy Lobb

J. WESTON
WALCH
PUBLISHER
PORTLAND, MAINE

User's Guide
to
Walch Reproducible Books

As part of our general effort to provide educational materials which are as practical and economical as possible, we have designated this publication a "reproducible book." The designation means that purchase of the book includes purchase of the right to limited reproduction of all pages on which this symbol appears:

Here is the basic Walch policy: We grant to individual purchasers of this book the right to make sufficient copies of reproducible pages for use by all students of a single teacher. This permission is limited to a single teacher, and does not apply to entire schools or school systems, so institutions purchasing the book should pass the permission on to a single teacher. Copying of the book or its parts for resale is prohibited.

Any questions regarding this policy or requests to purchase further reproduction rights should be addressed to:

Permissions Editor
J. Weston Walch, Publisher
321 Valley Street • P. O. Box 658
Portland, Maine 04104-0658

1 2 3 4 5 6 7 8 9 10

ISBN 0-8251-2855-2

Printed in the United States of America

CONTENTS

When the Earth Opens 1

To the Teacher

Almost every day, a natural disaster makes headlines somewhere in the world. The 1990's have had their share of tragedies caused by natural disasters.

- Hurricane Andrew hit Florida in August 1992, causing the greatest property damage of any natural disaster ever to hit the United States.
- The eruption of Mount Pinatubo in June 1991 in the Philippines was the largest volcanic eruption of the twentieth century.
- A cyclone followed by a storm surge roared up the Bay of Bengal toward Bangladesh in April 1991, killing at least 150,000 people in that flood-prone country.
- The Yangtze (Chang) River in central China flooded in 1995. Between May and July, 1200 people died in the extensive flooding.
- In Russia, an earthquake measuring 7.5 on the Richter scale occurred in May 1995. It flattened the town of Neftegorsk, killing thousands of people.
- An earthquake in northern Iran in June 1990 killed 36,000 people.

Other natural disasters appear less dramatic but are often even more deadly. In the African Sahel, millions have died from disease and starvation as a severe drought has lingered for years. Chronic droughts plague other areas of the world, including India, Australia, and the American West.

The number of deaths and the amount of property damage from natural disasters is on the rise around the world. The less-developed countries suffer 90 percent of the deaths from natural disasters, because of their huge populations and poor standards of living. Developed nations usually suffer fewer deaths but have enormous losses of property. Today, countries around the world are much more interdependent than they used to be. So, a disaster in one affects not just that country but others as well.

With all the scientific advances of the past century, we still have little ability to predict or control natural disasters. In fact, scientists have only a sketchy understanding of the causes of most natural disasters.

We may never fully understand the causes of all natural disasters. But we can improve our chances of surviving a natural disaster by being aware of those likely to occur where we live and by being prepared ahead of time for them.

Thus, the purpose of *Natural Disasters* is to explain what scientists believe to be the causes of a variety of natural disasters and to suggest steps you and your students can take to be better prepared in the event of a natural disaster.

How to Use This Book

Natural Disasters is a reproducible teacher book containing both student sections and teacher sections. Each teacher section includes vocabulary, an answer key, and information pertaining to the student section that follows it.

The student sections contain short reading selections followed by a Review page containing comprehension questions. Each Review includes factual recall questions as well as comprehension and application questions. You may wish to instruct students to write their answers to comprehension and application questions on the back of the sheet or on a separate piece of paper.

Three sections following each unit contain further information, suggested student activities, and a list of vocabulary words with definitions.

Reproducible pages are identified by the copyright line with a flame logo at the bottom. They can be copied and distributed to each student.

You can present all the units in *Natural Disasters,* or you can use only selected units, as time permits.

When the Earth Opens

Teacher's Guide for pp. 3–5

This teacher material covers the following reading selection:
Mount St. Helens: A Sleeping Giant Wakes Up

Vocabulary (See pp. 20–21 for definitions.)

avalanche
crater
evacuate
Hiroshima
seismographs

Answer Key (Review 1, p. 5)

1. Earthquakes began shaking the mountain, increasing in number and severity.
2. (any three) ash; heat; floods; mudflows; hurricane-like winds; falling trees
3. A restricted area was set up 10 miles around the peak. Residents of the area were asked to evacuate.

Answers for the following will vary.

4. They wanted to see the volcanic activity. They didn't really believe they were in danger. Reporters and scientists were trying to collect more information.
5. Yes. When people can actually see the damage that was done, they will be more likely to listen to warnings.
6. Yes. It is an area that will be under continual risk from eruptions.

Additional Information on Mount St. Helens

- Earlier eruptions of Mount St. Helens occurred in 1800, 1831, and 1835. The years from 1842 to 1857 are known as the Great Eruption period because of the many intermittent eruptions that occurred. These were witnessed by Native Americans and settlers.
- Native American tribes of the Columbia River each have a legend about the creation of Mount St. Helens. One tells the story of Loo-wit, a beautiful Indian maiden loved by two sons of the Great Spirit. When she could not choose between them, the two sons fought. The Great Spirit was angered and changed her into Mount St. Helens. The two sons were changed into Mount Adams and Mount Hood.
- Mount St. Helens is an example of a strato-volcano or composite volcano. The cone of the volcano is steep and symmetrical.
- Technology enabled volcanologists to predict the eruption of Mount St. Helens two weeks in advance. This was a first.
- The only other mountain recently active in the Cascades was Lassen Peak in California in 1914. Mount Hood erupted in the 1880's. Mount Baker erupted in 1843, 1854, and 1857–58. Mount Rainier steams periodically.

Mount St. Helens: A Sleeping Giant Wakes Up

Mount St. Helens is a spectacular peak in Washington state. It is in the Cascade Range, an area of forests and clear blue lakes. Each year, millions of people visit this scenic area.

Mount St. Helens is also the most active volcano in the Cascade Range. For 123 years, it had slept peacefully. It gave no clue to the fury hiding under its serene, snow-capped peaks.

But on March 20, 1980, Mount St. Helens woke up. On that day, a small earthquake shook the mountain. More small quakes took place the next day and the next.

Scientists knew these earthquakes were a sign that the volcano was coming to life. To keep a close watch on things, they set up seismographs to measure the earthquake activity.

The quakes became stronger. Soon they were setting off avalanches on the mountain slopes. But no signs of an eruption appeared.

A week after the first quake, a cloud of steam and ash burst out of the volcano. Similar eruptions continued over the next few days.

Some people who lived near the volcano began to evacuate their homes. Other people wanted to get front-row seats for the show. Sightseers, reporters, and scientists moved in closer. Many flew over the volcano for a bird's-eye view. On Sunday afternoon, March 30, 250 planes were seen cruising the volcano. Some people hiked up the mountainside. Others sold souvenirs to the many sightseers. In general, a party attitude prevailed.

Meanwhile, the quakes continued. The crater at the summit began to enlarge. A huge bulge on the north side of the volcano began to swell about 5 feet a day.

Governor Dixie Lee Ray restricted a 10-mile area around the peak to keep the volcano watchers out of the danger zone. Residents of Spirit Lake, directly below the mountain, were ordered to evacuate. Most did so willingly.

One who did not was 84-year-old Harry Truman. Truman, no relation to the president with the same name, ran St. Helens Lodge. He had lived most of his adult life on the mountain. He made it clear that he was not going to leave.

"I have some people ask me why I stay," Truman said. "That's my life—Spirit Lake and Mount St. Helens. I've lived here 50 years—it's a part of me. That mountain and that lake is a part of Truman. And I'm a part of it." So he stayed put.

On the morning of May 18, the fireworks began. A huge explosion occurred. It was heard hundreds of miles away. The blast was 500 times more powerful than that of the atom bomb dropped on Hiroshima in World War II.

Hurricane-like winds of more than 200 miles per hour felled trees for 230 square miles

(continued)

Mount St. Helens: A Sleeping Giant Wakes Up *(continued)*

around. Blizzards of ash shot 12 miles into the air. Winds carried the ash and dropped it over Oregon, Idaho, and Montana. The intense heat in streams near the volcano killed millions of fish.

The volcano's heat melted the snow. The melted snow mixed with the ash. The result was a hot, thick, smelly mud. The mud, at temperatures up to 211°F, raced down the Toutle River at 50 mph. It destroyed everything in its path.

In Yakima, Washington, 100 miles away, it was dark at noon as black clouds of ash blocked the sun. Ash fell 10 inches deep on the town.

The volcano roared all day and most of the night. It shot out millions of tons of gas and ash.

Harry Truman, in his beloved St. Helens Lodge, was buried under 40 feet of boiling mud. He was one of 63 people who lost their lives that day. Some of these people had refused to leave the area. Others thought that they were safe since they were outside the restricted zone.

The eruption continued for four days. It destroyed millions of dollars worth of timber, crops, homes, and other property. It turned beautiful forests within 20 miles of the volcano into wastelands. Cleanup costs were monumental.

Mount St. Helens will never be the same. The eruption tore almost 1300 feet off the top of the mountain. It left a horseshoe-shaped crater about 1.2 miles wide.

The mountain is quiet now. Once again, snow and ice cover its heights. Life near the mountain has gone on. People are logging the fallen trees. Others have begun farming again. And the Mount St. Helens area has been set aside as a national monument to educate people about volcanoes.

Name ______________________________

Date ______________________________

VOLCANOES
REVIEW 1

Get the Facts

1. How did scientists know that Mount St. Helens was coming to life after 123 years of quiet?

2. Name three destructive forces caused by an eruption.

3. What measures did officials take to keep people safe from the volcano?

What Do You Think?

4. Why do you think so many people flocked to see Mount St. Helens erupt, in spite of the danger?

5. Mount St. Helens National Monument was set up to educate people about volcanoes. Do you think this type of information will help people be more cautious around volcanoes? Why or why not?

6. Do you think the area around Mount St. Helens should be barred from development and saved only for recreational use? Why or why not?

Teacher's Guide for pp. 8–11

This teacher material covers the following reading selections:

How Are Volcanoes Formed?
What Happens When a Volcano Erupts?

Vocabulary (See pp. 20–21 for definitions.)

ash	crust	plug
bombs	lava	Ring of Fire
caldera	magma	shield volcano
carbon dioxide	mantle	stratovolcano
cinder cone	molten rock	symmetrical
core	mudflow	tsunami
crater	plate	vent

Answer Key (Review 2, p. 11)

1. crust; mantle; core
2. They rest on molten rock which moves about slowly.
3. at the edges of plates
4. When the plates pull apart, a crack is formed through which the magma can flow easily. No pressure builds up.
5. earthquakes; a bulge in the side of the volcano; emission of gases from cracks
6. Magma is the liquid rock in the mantle. It is called lava when it reaches the surface.
7. As magma comes out of the volcano, some of it cools and becomes solid. Small pieces are called ash.
8. sulfur fumes; carbon dioxide

Answers for the following will vary.

9. Large quantities of ash in the air would block the sun, causing a drop in temperature. If large amounts of steam are emitted, rainstorms may be produced.

Additional Information on Volcanic Eruptions

- Volcanoes may be in one of three stages:
  - (a) Active volcanoes have frequent eruptions. There are about 600 active volcanoes in the world today.
  - (b) Dormant volcanoes have erupted in the past, but have not done so for a long time. However, it is possible they will erupt again.
  - (c) Extinct volcanoes have shown no activity for so long that scientists feel it is unlikely they will ever erupt again.
- There are five different types of volcanic eruptions. Here they are, in order of the violence of their eruptions, from gentlest to most violent:
  - (a) Hawaiian volcanoes have the gentlest eruptions. Steady, slow eruptions of red-hot, runny lava occur. Gas can escape easily without explosions.

    Example: Kilauea, in Hawaii
  - (b) Strombolian volcanoes have frequent small eruptions of thicker lava. Small explosions may occur every few minutes. Eruptions may continue over months or years.

    Example: Stromboli, island off Italy
  - (c) Vulcanian volcanoes blast solid blocks clear of the volcano. They produce much ash and gas. The explosions caused by escaping gas may blow away parts of the volcano.

    Example: Vesuvius in Italy
  - (d) Plinian volcanoes erupt with a huge number of bombs and ash. Their eruptions often have enough force to destroy large parts of the volcano. They were named after the scholar Pliny the

Elder, who died in the eruption of Mount Vesuvius in A.D. 79.

Example: Mount St. Helens in 1980

(e) Peléean eruptions are the most violent. They involve very thick lava. The lava often forms a plug in the vent of the volcano. Violent gas explosions remove the plug. The explosions throw out glowing gas clouds. This type of eruption was named after Mount Pelée, which buried the town of Saint-Pierre, West Indies, in 1902.

◆ Alfred Wegener proposed the theory of continental drift in 1912. He believed that the continents were once joined as one giant landmass called Pangaea. The landmass split and began drifting apart 200 million years ago. This idea created an uproar for 50 years. But today, most scientists have accepted the theory of continental drift.

The following evidence supports this theory.

(a) Fossils from identical extinct animals have been found on widely separated continents. This is called the fossil link.

(b) Rocks also match up. Folded mountain ranges, coal deposits, and glacial deposits line up perfectly when South America and Africa are joined.

(c) The theory of plate tectonics states that the earth's top layer, or crust, is made up of seven major plates. Inner currents move these plates around on the earth's middle layer, or mantle.

(d) Ocean-floor spreading occurs along deep cracks that run through the center of under-ocean mountains. These cracks, or rift valleys, are found in mid-ocean ranges in all oceans. Lava erupts from the rift valley. The lava hardens, pushing away the ocean floor on either side of the ridge and forming new ocean floor.

(e) In subduction, older rocks sink into the interior of the earth. New rocks, formed along the mid-ocean ridges, push the older rocks into deep trenches at the bottom of the ocean. Subduction and ocean-floor spreading balance each other. Thus, the earth's crust remains the same size.

◆ When the earth was formed about four billion years ago, it was very hot. Over time, the crust has cooled. But the interior is still very hot. The heat inside the earth may be maintained by radioactive materials in the rocks. As these materials break down, they produce a small amount of heat. This heat builds up in the inner layers of the earth. The radioactive substances are probably gradually dwindling away. But the interior of the earth should remain hot for billions of years to come.

◆ Some volcanoes are not formed at the edge of plates. They are caused by hot spots in the mantle. Molten rock in a hot spot may melt through the crust. An example of this is Kilauea volcano in Hawaii.

◆ A large number of the world's volcanoes lie under the ocean. When these build themselves up above sea level, lava flows may form islands. An example of a new island formed this way is Surtsey, Iceland, which was created in 1963.

There are 10 times as many active volcanoes under the ocean as there are on land. Almost the entire ocean floor is covered with basalt, which is hardened lava.

◆ Underground molten rock and gas cause hot springs and geysers. The gas seeps through the crust and heats up water in the ground. Geysers shoot hot water and steam into the air. There are geysers in Yellowstone National Park in Wyoming, Iceland, New Zealand, Japan, and elsewhere.

In some areas, the hot water and steam are produced very gently. In this case, hot springs or mud pools are formed instead of geysers. In Iceland this geothermal energy is used to heat homes and to drive turbines that produce electricity.

How Are Volcanoes Formed?

The earth is made up of three layers: the crust, the mantle, and the core. The crust is the land on which we live and the land beneath the oceans. On the average, the crust is about 50 miles thick. Under the crust is a layer called the mantle. Under the mantle is the core.

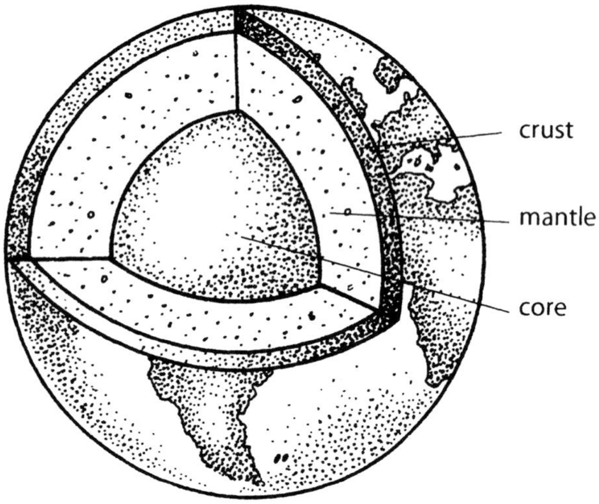

Earth's three layers

The temperature of the earth rises as we go down toward the core. It is thought to be about 7200°F or more in the core. Even so, most of the earth is solid because of the great pressure, which increases with depth.

At the upper part of the mantle, it is hot enough and the pressure is low enough that some of the rock melts. This molten rock is called magma. As more magma forms, it rises toward the surface through cracks in the crust.

The earth's crust is not one solid piece. Rather, it is made up of huge pieces called plates. The plates fit together much like a jigsaw puzzle. These plates move about slowly on the molten rock of the mantle on which they rest. You can't see this movement. But scientists know it is there by comparing pictures taken by satellites over time.

There are seven major plates. As they float on the mantle, they bump together or pull apart. In places where the plates pull apart, large cracks open in the crust. Then magma flows freely to the surface. If it reaches the surface, it is called lava.

This free-flowing lava produces a broad, gently sloping volcano that looks a little like a shield, called a shield volcano. Shield volcanoes do not usually erupt explosively. They simply ooze large amounts of lava continuously. Shield volcanoes may damage property. But they rarely cause loss of life. They are usually found on mid-ocean islands, such as Iceland or Hawaii. An example is Mauna Loa in Hawaii.

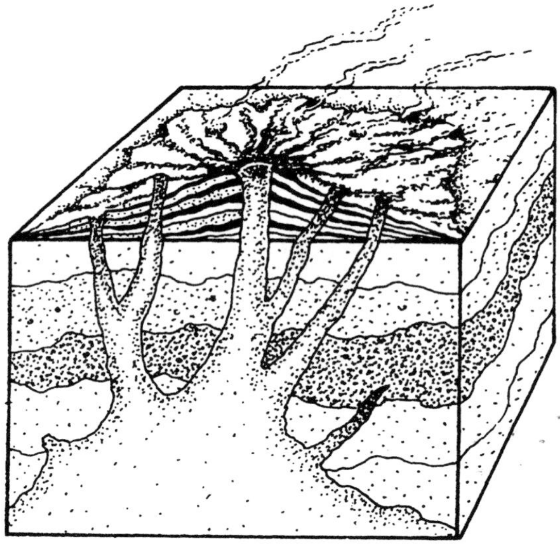

Shield volcano

More violent volcanic eruptions are caused when two plates collide and one plate sinks under the other. Rocks on the lower plate melt and form magma. The magma pushes its way up through the crust through openings called vents. Stratovolcanoes and cinder cones are formed in this way.

(continued)

How Are Volcanoes Formed? *(continued)*

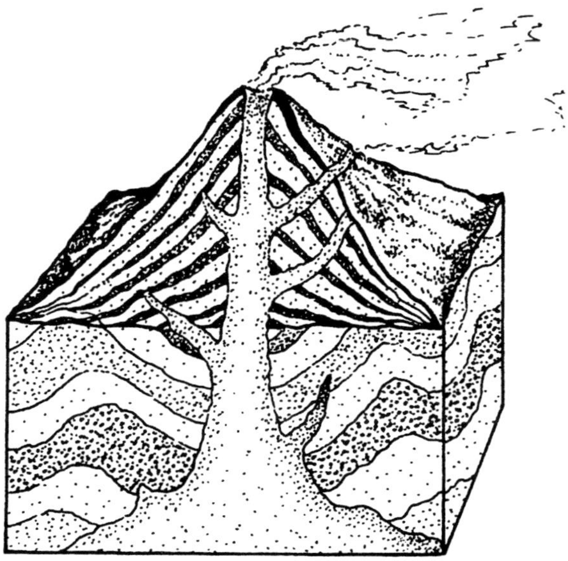

Stratovolcano

With a stratovolcano, thick lava erupts from the vent. It flows slowly, usually solidifying before it goes very far. After a few eruptions, a small hill of lava builds up around the vent. More layers are deposited with each eruption. The volcano grows into a tall, cone-shaped mountain. These volcanoes are called stratovolcanoes because they are formed in layers (strata). An example is Mount St. Helens.

A third type of volcano is called a cinder cone. These are formed when a volcanic eruption shoots out mostly solid material rather than liquid lava. This results in a pile of loose material around the vent of the volcano. An example of a cinder cone is Paricutín, in Mexico.

All three types of volcanoes nearly always occur at the edges of the plates. So volcanoes are found around the earth in long thin lines. One of these plate edges circles the Pacific Ocean. Because of the many explosive volcanoes on this line, it is known as the Ring of Fire. Eighty percent of the earthquake and volcanic activity in the world occurs around this ring.

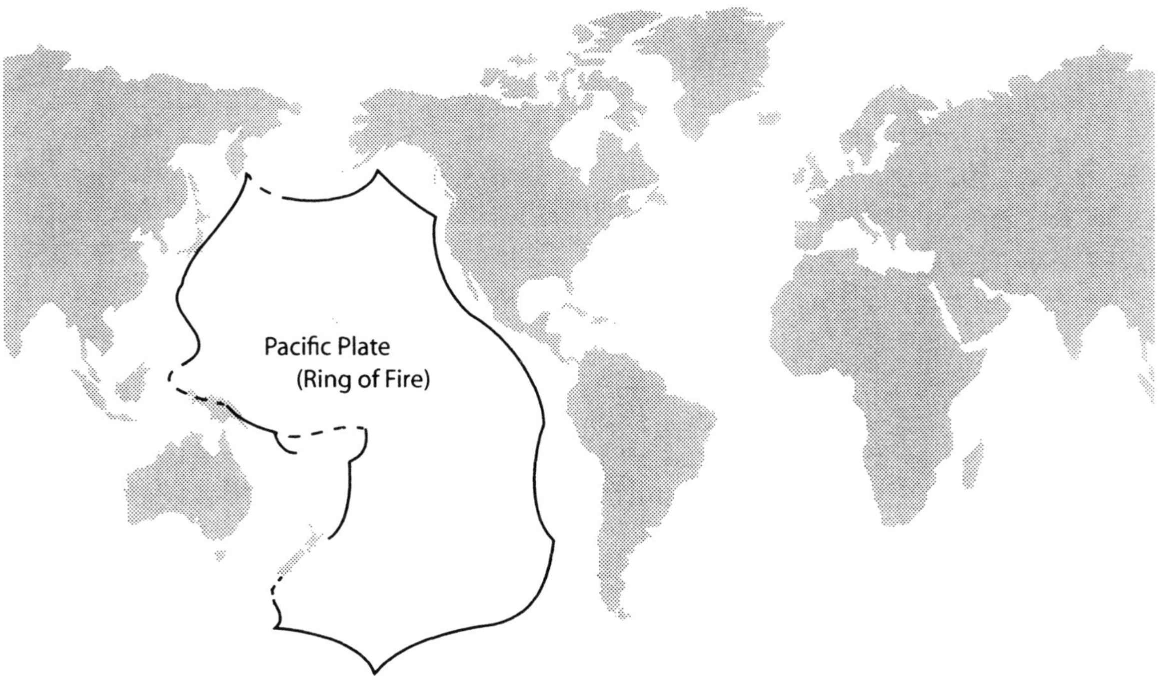

What Happens When a Volcano Erupts?

Have you ever shaken a bottle of soft drink? When you take the lid off, it blows sky high. Volcanoes act in a similar way. The vent of the volcano is blocked with a plug of rock. The plug keeps the magma in its place beneath the ground. If enough pressure builds up inside the volcano, the plug will be blown out and the volcano will erupt.

Usually there are signals that a volcano is about to erupt. The ground may begin to shake. The sides of the volcano may bulge as the hot magma gathers under the surface. Gases may stream out of cracks in the rock. All these signals may go on for weeks, months, or longer. Or . . . there may be no warning at all.

When a volcano does erupt, it makes an awesome noise that may be heard for hundreds or thousands of miles. Liquid rock, gases, steam, and dust blast out of the vent. No living thing nearby has time to escape the fury.

After the initial blast, magma from deep within the earth erupts from the volcano. The lava flows out, destroying everything in its path. It is extremely hot—from 1470° to 2100°F.

As the magma comes out of the volcano, some of it may cool and become solid. These solid pieces shoot out of the volcano with terrific force. The smallest pieces are called ash. Larger pieces are called bombs. Bombs are usually between the size of a tennis ball and a football. But they may be much larger—up to 3 feet across and weighing 100 tons.

Sections of the volcano itself may explode into the air. When this happens, a large pit called a crater may be formed at the top of the vent. The largest craters, called calderas, are formed when the entire top of the volcano collapses inward.

The volcano blows off poisonous gases such as sulfur fumes and carbon dioxide. These fumes can kill people and wildlife in the area.

The volcano also emits steam. Water seeps into the volcanic rock from the sea or a river. When the water reaches the magma, it turns to steam. As the steam is blown into the air, it cools and forms water drops. If there is a lot of steam, there may be a rainstorm. And when the rain mixes with the ash that has just fallen, a thick, black mud is formed. Hot mudflows rush along at high speeds, destroying all they touch.

Volcanic eruptions in the ocean may cause tsunamis—giant waves. When Krakatoa erupted in 1883 on an island in the East Indies, the tsunamis it created washed 36,000 people out to sea.

A volcano may continue erupting for days, months, or even years. Then it may stop. If this happens, chances are that the vent has become blocked with lava that has hardened in it, forming a plug. Over time, the vent will fill in. The volcano will then appear to be just another mountain. Maybe it is. But maybe when conditions are right, it will erupt again.

Name ______________________________

Date ______________________________

VOLCANOES
REVIEW 2

Get the Facts

1. Name the three layers of the earth.

2. Why do the pieces of the earth's crust move about?

3. Where do volcanoes almost always form?

4. Why are those volcanoes that are made where plates have pulled apart less dangerous than those made by colliding plates?

5. What are some signals that a volcano is about to erupt?

6. What is the difference between magma and lava?

7. How is volcanic ash formed?

8. Name two poisonous gases volcanoes give off.

What Do You Think?

9. In what ways might a large volcanic eruption affect the weather in the surrounding area?

Volcanoes

Teacher's Guide for pp. 13–15

This teacher material covers the following reading selections:

What Happens After the Eruption?
What Is a Volcano Good For?
Can Eruptions Be Predicted or Controlled?

Vocabulary (See pp. 20–21 for definitions.)

dormant	seismographs
geothermal energy	tiltmeters
infrared photography	volcanologist
pioneers	

Answer Key (Review 3, p. 15)

1. mosses and lichens
2. (1) mosses (2) pioneer plants (3) butterflies (4) deer
3. potassium; phosphorus
4. The heat of the eruption liquefies the metals in the rocks. The metals flow up the cracks in the crust. Later, they cool and solidify into deposits.
5. heating homes; running turbines to make electricity
6. It measures swelling of the volcano.
7. It detects earthquakes and smaller shocks.
8. not to live too close to them
9. Reflectors are placed around the volcano. Laser lights are aimed at the reflectors. The time it takes for the light to bounce back is measured. Shorter time indicates the volcano has swollen.

Answers for the following will vary.

10. Scientists would like to drill into a magma chamber and extract its heat. The problem is that currently there are no materials for making drills that could survive such high heats. However, natural hot water and steam in geyser regions have been used to provide hot water and power.

What Happens After the Eruption?

Finally, the eruption is over. The volcano is once again in a dormant period. Now the land around the volcano is barren as far as the eye can see. The only sign that life once existed here is the remains of dead trees scattered about.

But soon, new life will begin on the harsh rocks. The first plants to grow are usually mosses and lichens. They spring up in cracks between the volcanic rock. Then they begin to break down the rock. That process forms soil in which other plants can grow.

The wind blows seeds of other plants into the area. Some of the seeds grow among the rocks. They form the first plants, which are called pioneers. After the plants come the insects, such as butterflies and beetles. The insects feed on the plants and their pollen. Soon after, come the animals. They feed on the insects and the plants.

What Is a Volcano Good For?

In the wake of a volcanic disaster, it may be hard to imagine any good coming from an eruption. But there are benefits.

Formation of Rich Soil

Volcanic ash breaks down quickly. As the ash breaks down, it releases potassium and phosphorus. These minerals fertilize the surrounding soil. Volcanic rock takes years to break down. But when it does, it forms a rich soil, full of nutrients needed to grow crops. Farming is good in areas around volcanoes.

Mineral Deposits

The heat of a volcanic eruption causes metals in nearby rocks to become liquid. The liquid metals flow up cracks in the rocks. Later, they cool and solidify.

Deposits of gold, silver, copper, tin, and iron have been formed in this way. These deposits may be discovered when erosion wears some of the rock away or when people dig for precious metals near old volcanoes.

Geothermal Energy

Sometimes hot springs, mud pools, and geysers form near volcanic areas. First, water runs deep into the earth through volcanic cracks. Then, the water reaches the hot rocks far beneath the surface and is heated. Finally, incoming cold water pushes the hot water to the surface as hot springs or geysers.

People living near the area can use this hot water to heat their homes. Power plants can use the steam to run turbines to produce electricity.

Scientific Knowledge

The study of volcanoes has led to new understanding of what lies in the interior of the earth.

Beauty

Volcanoes provide some of the most beautiful scenes in nature. Each year, millions of people flock to see the wonders of the Cascade Range, Yellowstone National Park, and other volcanic areas in the western United States.

Can Eruptions Be Predicted or Controlled?

Millions of people around the world live near volcanoes. An unexpected eruption could endanger whole cities. So scientists have developed ways to keep track of volcanic activity.

Tiltmeters

As magma collects in pockets inside the volcano, the volcano may swell in that area. The swelling may be too slight to be seen by the eye. But instruments called tiltmeters are sensitive enough to detect small changes. By placing a number of these around the volcano, scientists can watch for any changes.

Seismographs

Before an eruption, magma begins to move inside the volcano. This movement sets off small earthquakes. The seismograph records these shocks and measures their strength. On a quiet volcano, perhaps five small shocks might be detected on a normal day. But just before an eruption, 600 or more might be measured.

Infrared Photography

The temperature of a volcano rises before an eruption. Infrared photography shows that the ground is heating up.

Laser-Reflection Systems

Lasers are used to measure any enlargement of the crater. Laser lights are aimed at reflectors placed around the crater. Changes in the time required for the light to be reflected back show how much the crater is swelling.

All these techniques can help scientists tell when a volcano is waking up. However, there is no way to tell exactly when the volcano will erupt or what it will do when it erupts. The problem is that not all volcanoes follow the same pattern. Scientists called volcanologists study each volcano individually to learn its ways. For this purpose, research stations have been built on the slopes of many volcanoes.

Once a volcano begins to erupt, there is nothing anyone can do to stop it. In the case of very slow lava flows, dams can be built to try to control the course of the flow.

The best way to keep safe from volcanoes is to not live too close to them. Many people think that since a volcano has not erupted in their lifetime, it poses no danger. But this may not always be true.

Name ______________________________

Date ______________________________

VOLCANOES
REVIEW 3

Get the Facts

1. What are usually the first plants to grow on volcanic rock?

2. Number each sentence from 1 to 4 showing the order in which life returns to a volcanic area.

   _____ pioneer plants grow from seed

   _____ large animals like deer

   _____ mosses

   _____ butterflies and other insects

3. Name two products of volcanic ash which fertilize soil.

4. Describe how volcanic eruptions form deposits of metals.

5. Name two uses of geothermal power.

6. What is the purpose of a tiltmeter?

7. What is a seismograph used for?

8. What is the best way to keep safe from volcanoes?

9. How do laser-reflection systems tell scientists how much a volcano is swelling?

What Do You Think?

10. How do you think volcanic energy might be harnessed to provide power? What would be the biggest problem in doing this?

Additional Information

America's Volcanic Landmarks

- Yellowstone National Park in Wyoming has more geysers than all the rest of the world put together. Geysers are formed when water seeps down to the magma and is turned to steam. The steam rises and comes out of the earth as a geyser. Although there have been no eruptions in Yellowstone for thousands of years, the geysers are a sign of volcanic activity.
- The Cascade Range in Washington, Oregon, and California contains many active volcanoes.
- Crater Lake in Oregon is the deepest lake in the United States. It was formed when the crater of an extinct volcano filled with water.
- Craters of the Moon National Monument in Idaho consist of fields of lava.
- Mount St. Helens National Volcanic Monument in Washington state was set up to educate people about volcanoes (see pp. 2–5).
- Ship Rock in New Mexico is a lava plug that is all that is left of a volcano.
- The petrified forest in Arizona shows the remains of trees that were uprooted during a volcanic blast millions of years ago.
- There are many active volcanoes in Alaska and Hawaii.

Volcanoes Throughout History

The Romans believed that Vulcan, the god of fire, worked in his forge under an active volcano in Sicily. They named the mountain Volcano after the fire god.

In the Middle Ages, people thought of volcanoes as gateways to hell.

Natives of Hawaii feared Pele, the goddess of the volcano. They made sacrifices to her to appease her anger.

Famous Volcanic Eruptions

Vesuvius

In A.D. 79, Mount Vesuvius in Italy erupted near the towns of Pompeii and Herculaneum. A huge surge of rock, ash and steam rushed down the mountainside at 60 mph. It killed people where they stood and buried them under 10 feet of rubble. Centuries later, the towns were uncovered. Historians found perfectly preserved artifacts and people. The discovery provided a clear record of life in those times.

Krakatoa

On August 27, 1883, Krakatoa blew itself to bits. The noise was heard 3000 miles away. Some 36,000 people were killed in the tsunamis that followed.

Mount St. Helens

See pages 2–5.

Paricutín

This volcano was born in a Mexican cornfield in 1943. Farmer Dionisio Pulido was working in his fields when a fracture ripped open the earth.

(continued)

Additional Information *(continued)*

Smoke and ash shot from the fracture. By the next morning, a cinder cone 30 feet high had formed. The volcano continued to grow over the next 10 years to a height of 1200 feet.

Mount Pelée

In 1902, Mount Pelée on the Caribbean island of Martinique erupted. It sent an avalanche of gases and ash down on the town of Saint Pierre. This 1450°F avalanche moved with hurricane speed. It claimed 30,000 lives in seconds. Only two townspeople survived.

Nevado del Ruiz

In 1985, Nevado del Ruiz in the Andes mountains of Colombia erupted. The eruption melted huge amounts of snow. It caused one of the biggest mudflows in history. And it killed 23,000 people in the town of Armero.

Mount Etna

The 1669 eruption of Mount Etna in Sicily killed 20,000 people.

Mount Toba

When Mount Toba erupted 600,000 years ago, all the ash and dust it threw into the air may have blocked the sun long enough to cause the last Ice Age.

Volcano Statistics

Big and Tall Volcanoes

The largest volcano is Mauna Loa in Hawaii. Its base stretches across 74 miles. It rises 33,000 feet from the ocean floor. But only about 14,000 feet of it is visible above the water.

The largest caldera is that of Mount Aso in Japan. It measures 17 by 10 miles.

The tallest extinct volcano is Cerro Aconcagua in Argentina. It stands 22,835 feet high.

The tallest active volcano is Ojos del Salado in the Andes of South America at 22,588 feet.

The tallest dormant volcano is Llullaillaco in the Andes of South America at 22,109 feet.

The tallest volcano known is Olympus Mons, an extinct volcano on the planet Mars. It is 15 miles tall and 370 miles in diameter.

The longest lava flow in historic time is from the 1783 eruption of Laki in Iceland. It ran 40 to 43 miles. Some 15 million years ago, a 300-mile lava flow oozed across North America.

Big Explosions

The loudest explosion was that of Krakatoa in 1883. It was heard 3000 miles away.

The most violent eruption was the eruption of Tambora in Indonesia in 1815. It left a caldera 7 miles across. It also discharged the largest amount of volcanic material—about 36 cubic miles of it.

Recent Volcanoes

The island of Surtsey off the coast of Iceland was created by a volcano in 1963.

Greatest Loss of Life

When Mount Pelée erupted in 1902, between 30,000 and 40,000 people died in the town of Saint-Pierre in Martinique.

The eruption of Krakatoa in 1873 resulted in a huge tsunami which killed 36,000 people.

The Tambora eruption in Indonesia in 1815 killed 92,000 people.

(continued)

Additional Information *(continued)*

Most Eruptions

Some volcanoes erupt almost all the time. Stromboli in Italy, Yasur in the Pacific, and Erta Ale in Africa have been erupting for hundreds of years.

Record-Holding Geysers

Steamboat Springs geyser in Yellowstone National Park squirts about 400 feet, to earn the record of highest-spurting geyser.

Old Faithful in Yellowstone is the most consistent geyser, spurting about once an hour.

Other Facts

The only continent with no volcanoes is Australia.

There are volcanoes on Antarctica. Big Ben erupted in 1960, Deception Island Volcano in 1970, and Mount Erebus in 1991.

Volcanoes on Other Planets

Space probes have discovered volcanoes on other planets. Mars has a group of four huge extinct volcanoes. Olympus Mons is the tallest. It is nearly three times the height of Mount Everest.

Venus also has volcanoes created by hot spots under its crust. Maxwell is the tallest. It stands a mile higher than Mount Everest. It is thought to be extinct. Other volcanoes on Venus, such as Rhea and Theia, erupt continuously.

Io, one of Jupiter's moons, has spectacular volcanoes that erupt continuously. *Voyager* space probes took pictures of some of them: Pele, Prometheus, and Loki.

Our own moon has many dome-shaped extinct volcanoes as well as old lava flows.

Volcanology: The Science of Volcanoes

Volcanologists study volcanoes. Their study helps them to predict eruptions and to understand how a volcano works. Often these scientists work in observatories close to the volcanoes.

They use special equipment to keep track of what the volcanoes are doing. They use seismographs to measure tremors within the earth. They use tiltmeters to measure the expansion of volcanoes.

Sometimes volcanologists go into a volcano for a closer look. If the volcano is active, they wear a suit of special aluminum material which protects them from the heat. The suit enables them to get within 3 feet of a lava flow. If they stepped into the flow, the lava would burn through the suit. They also wear a helmet to shield the head from falling particles. And they wear a mask for protection from poisonous gases.

Activities for Students

1. Examine the theory of continental drift. Use tracing paper to trace over the outlines of the continents on a world map. Cut the continents out. Try to put them together like a puzzle. Do they fit?
2. Project Mohole was a study aimed at drilling a hole through the earth's crust to the mantle. Find out how deep scientists were able to dig during this project.
3. Look in the newspaper for articles about volcanoes. Locate on the globe any volcanoes mentioned. Do the volcanoes lie on the edge of a plate?
4. Make a model of a volcanic eruption. Get an empty bottle with a narrow neck. Put the bottle in a sink or basin. Pour 1 cup of warm water into the bottle. Add a few drops of liquid soap to the water. Add 1 tablespoon of baking soda to the warm, soapy water. If a reaction does not begin, add 1 tablespoon of vinegar.

   The eruption occurs because when you add baking soda to the water, carbon dioxide is released. The gas expands and rises to the top. In a volcano, gases in the magma expand and force the magma out of the volcano.
5. Imagine that you are a reporter covering the eruption of a volcano. You are on the scene. The volcano begins to erupt. Prepare a news flash that will interrupt local programming to announce the eruption.
6. Write a paragraph describing the destruction of wildlife and habitat by an eruption such as that of Mount St. Helens.
7. Create a TV program in which you tell first-person stories of people who survived a volcanic eruption.
8. Make a prediction about when Mount St. Helens might erupt again. Tell what your prediction is based on.
9. Find out more information about Hawaii's major volcano, Mauna Loa.

10. Find out more about careers in volcanology.
11. Find out more about Yellowstone National Park. Write to the park for brochures.
12. Read about geysers in New Zealand.
13. Find out about the United States Geological Survey.

Volcano Vocabulary

active volcano	volcano that erupts from time to time
ash	small particles of lava, less than 0.16 inch across, which are shot out of a volcano
avalanche	material rushing down the sides of a mountain
bomb	larger piece of lava thrown out of a volcano, which starts as a molten blob and hardens as it flies through the air
caldera	large hollow area at the top of a volcano formed when the sides of the volcano collapse inward
carbon dioxide	gas that is invisible, odorless, and suffocating if breathed in large quantities
cinder cone	pile of ashes and bombs around a volcanic vent
crater	opening at the top of a volcano caused by the force of the explosion; smaller than a caldera
core	inner layer of the earth
crust	thin, outer layer of the earth
dormant volcano	volcano that has not erupted for some time but that may do so again in the future
eruption	explosion of lava, ash, and gases from a volcano
extinct volcano	volcano that is dead and will not erupt again
evacuate	move people out of a danger zone
geothermal energy	power made from the heat of the inner earth
geyser	fountain of hot water and steam erupting from the ground in a volcanic area
Hiroshima	Japanese city on which an atomic bomb was dropped in World War II
igneous rock	rock formed from molten rock (magma and lava)
infrared	type of photography that measures hot areas
lava	magma that has been forced to the surface during a volcanic eruption
magma	molten rock beneath the earth's crust, known as lava when it reaches the surface
mantle	middle layer of the earth, between the crust and the core

(continued)

Volcano Vocabulary *(continued)*

molten rock	rock in a liquid state (magma or lava)
mudflow	mixture of water and ash
Ring of Fire	ring of active volcanoes around the edge of the Pacific plate
pioneer plants	first plants to grow in a barren area
plate	section of the earth's crust
plug	mass of solid lava that blocks a volcano's vent
seismograph	instrument for detecting earthquakes
shield volcano	volcano with a broad base and rounded sides
stratovolcano	volcano with steep, symmetrical sides
symmetrical	even or balanced in shape
tiltmeter	instrument that detects the swelling of a volcano before an eruption
tsunami	huge ocean wave caused by volcanic eruption or earthquake under the sea; incorrectly called a tidal wave
vent	vertical pipe in a volcano through which the magma rises
volcanologist	a scientist who studies volcanoes

Teacher's Guide for pp. 23–25

This teacher material covers the following reading selection:
The Good Friday Earthquake

Vocabulary (See page 42 for definitions.)

avalanche
tsunami
Richter scale

Answer Key (Review 1, p. 25)

1. 1964
2. Part of the cliff they were located on broke off and fell into the sea.
3. fire and high waves
4. tsunami
5. tsunami
6. Alaska is sparsely populated, and it was a holiday so people were not at school or work.
7. (any two) strongest quake in North America on the Richter scale (8.5); longest-lasting quake (7 minutes); greatest uplift of land by an earthquake (50 feet)

Answers for the following will vary.

8. They probably did not believe that an earthquake so far away could cause them harm. They may not have realized the strength of a tsunami.
9. probably so, if the area is obviously unstable

Additional Information on the Good Friday Earthquake

- Alaska is in the Ring of Fire—the great earthquake-prone band that runs around the Pacific Ocean.
- The International Tsunami Warning Center was set up in 1948 in Honolulu, Hawaii.
- The Good Friday quake was carefully studied by architects and scientists. It was found that homes built on bedrock will be more earthquake resistant. The houses at Turnagain Heights were built on clay, which was less solid ground. Buildings, dams, and power plants should be built with earthquake resistance in mind.

The Good Friday Earthquake

It was 5:35 in the evening on March 27, 1964. All was quiet in Anchorage, Alaska. It was a holiday. It was both Good Friday and the eve of Passover.

The schools were empty. Businesses had closed. The sky was dark. Most people were home for the night.

One exception was the downtown movie theater. It was playing a new Walt Disney film to a packed house.

One minute later, the whole world seemed to fall apart. First, there was a deep rumble that sounded a little like thunder. Next, there was a slight trembling. Then the earthquake became stronger and more violent.

Some buildings collapsed. Others sank into holes that opened beneath them. Cars bounced about. Large cracks opened in the streets.

In the movie theater, the lights went out. The audience of mostly children began to cry and scream. Fortunately, the manager and ushers stayed calm. They switched on the emergency lights and got the children out safely.

It was lucky the children were not in school at the time. Government Hill Elementary split in two when the ground beneath half of it dropped 10 feet.

The earthquake continued for four minutes. When it was over, much of downtown Anchorage was destroyed. The left half of 4th Avenue dropped 20 feet below the right half. In the next three days, some 300 aftershocks shook those buildings that were left standing.

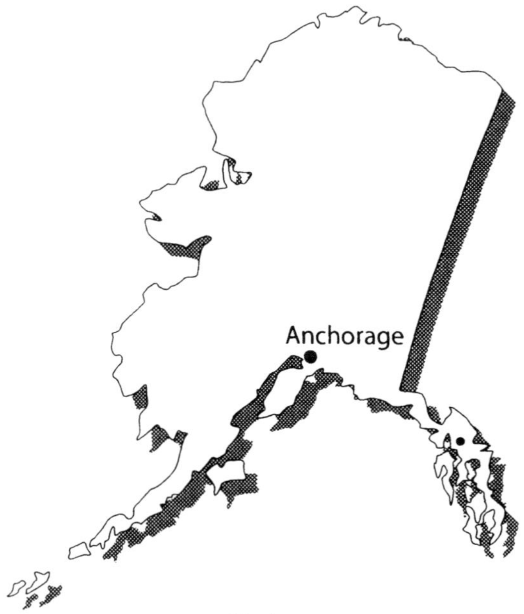

Alaska

In a suburb of Anchorage called Turnagain Heights, expensive houses had been built on a cliff overlooking Cook Inlet. When the quake struck, a mile-long section of the cliff broke off and fell into the sea. Houses, cars, and people went, too. What was left of the suburb sank 6 feet.

Anchorage was not the only place hit. There was damage across much of Alaska. Highways were ripped apart. Railroad tracks were twisted and tangled. Avalanches of ice and rocks rushed down mountainsides.

Huge amounts of oil were being stored in Seward, a port town. Oil was shipped into Seward by rail, held in tanks, and then sent out on ships. The quake hit. The tanks broke. And the oil caught fire. The entire waterfront

(continued)

The Good Friday Earthquake *(continued)*

collapsed into the bay. The flaming oil swept back into town. The fire spread. Much of the town was destroyed.

In Valdez, a small coastal town, buildings fell and streets cracked. But the main destructive force came from the sea. The earthquake caused huge waves, tsunamis. The waves crashed into town. Only a few people were able to reach safety. The town was destroyed. Months later, residents rebuilt the town in a safer location a few miles away.

At first, the damage in Kodiak, a coastal town located about 250 miles from Anchorage, did not seem too severe. But authorities feared tsunamis. They asked residents to leave for higher ground. Many left. Others stayed. The tsunamis came. They kept coming for most of the night. By the time they ended, much of Kodiak was destroyed. The downtown area was a jumbled mass of boats, fallen buildings, and smashed cars. But only eight people in Kodiak were killed.

In fact, the entire death toll from the quake was low: 115 people. That was probably because Alaska is so sparsely populated. Also, the fact that the quake occurred on the evening of a holiday meant that schools and businesses were closed. But the property damage was estimated at $750 million. The Good Friday quake was the strongest quake ever to hit North America, registering 8.5 on the Richter scale.

The quake changed the landscape of Alaska. A part of the Gulf of Alaska the size of the state of Maine rose 3 to 8 feet. The floor of the ocean between Kodiak and Montague islands rose more than 50 feet. That is the greatest uplift of land by an earthquake ever recorded. Another large area in the gulf sank. One island rose 38 feet. The Kenai Peninsula moved sideways about 60 feet.

The Good Friday Earthquake was felt around the world. Seismographs in England and Sweden registered the shocks.

Tsunamis caused by the quake traveled for thousands of miles at sea. In Crescent City, California, 1400 miles from the quake, a tsunami swept over the town, drowning 10 people. Tsunamis traveled at 450 mph across the Pacific reaching Hawaii, Japan, and even Chile, 8400 miles away.

The Good Friday quake was a record breaker. It was the longest-lasting earthquake ever. It began with 40 seconds of tremors. It continued with four minutes of hard shaking. The entire quake lasted seven minutes. And aftershocks continued for nine months. It was an earthquake of incredible force. No one who lived through it will ever forget it.

Name ______________________________

Date ______________________________

EARTHQUAKES
REVIEW 1

Get the Facts

1. In what year did the Good Friday quake occur?

2. Why were many homes in Turnagain Heights destroyed?

3. What caused the destruction of Seward, Alaska?

4. What force destroyed Valdez, Alaska?

5. How was Kodiak, Alaska, destroyed?

6. Why was the death toll from the quake so low?

7. Name two records set by the Good Friday quake.

What Do You Think?

8. Ten people in Crescent City, California, were drowned by a tsunami caused by the Alaskan earthquake. The town had adequate warning to evacuate. Why do you think many people did not do so?

9. In Turnagain Heights, many expensive homes were built on the edge of a cliff overlooking the water. They were destroyed by the quake. Do you think people should be restricted from building in such locations? Why or why not?

Teacher's Guide for pp. 28–33

This teacher material covers the following reading selections:

What Causes Earthquakes?
Where Do Most Earthquakes Occur?
Earthquake Faults
What Are the Effects of Earthquakes?
How Are Earthquakes Measured?

Vocabulary (See page 42 for definitions.)

aftershock
earthquake
epicenter
fault
focus
Mercalli scale
mantle
plates
primary waves (P waves)
Richter scale
secondary waves (shear waves or S waves)
seismic waves
seismograph
surface waves (L waves)
tsunami

Answer Key (Review 2, p. 33)

1. When two plates collide, great stresses are built up. One plate gives way with a jolt, causing an earthquake.
2. The epicenter is the spot on the earth's surface directly above the focus of the earthquake.
3. L wave or surface wave
4. the Ring of Fire, an area circling the Pacific Ocean
5. a crack in the earth's surface where two plates meet
6. San Andreas Fault
7. The Richter scale uses scientific instruments to measure the amount of energy released by the quake. The Mercalli scale is a subjective measure of the damage done by a quake.
8. 7.0 or higher (not exceeding 8.0)
9. (any 4) shocks, aftershocks, landslides, avalanches, fires, tsunamis

Answers for the following will vary.

10. If the epicenter of the quake is in a heavily populated area, it is much more likely to cause great loss of life. An earthquake in a remote area will probably cause much less damage. For example, the 1964 quake in Alaska was the strongest quake in North American history. It caused great property damage, but just 115 deaths. The death toll would have been much greater in a heavily populated area.

Additional Information on Earthquakes

- For many years it was believed that faults caused earthquakes. Modern technology, especially computers, has enabled scientists to determine that earthquakes originate in the earth's mantle. Faults are cracks in the earth's crust caused by the movements of the mantle.
- Every 30 seconds, somewhere around the world the earth shakes. Most of these earthquakes are too small to be noticed by people. But a few hundred each year are large enough to cause damage. Some 80 percent of all quakes occur in the Pacific Ring of Fire. Another earthquake belt runs from the Mediterranean area east across Asia.
- Primary waves are also known as longitudinal waves, push waves, or P waves. This kind of wave advances by pushing and pulling particles in its path along the direction of the wave's travel.

Earthquakes

- Secondary waves are also called shear waves, transverse waves, shake waves, or S waves.
- Surface waves are also called L waves. They are called this because they are the longest waves.
- When an earthquake occurs, both P and S waves are sent out in all directions from the focus of the quake. The P wave is faster and reaches distant points first. That is where it got its name (primary). The S waves arrive later and are therefore called secondary waves.
- P waves travel in solids, liquids, or gases. S waves travel in solids. L waves travel only on the earth's crust. This fact gives scientists a tool to discover information about the structure of the earth's interior. It shows that the earth is made up of layers of different kinds of materials. It can also show likely places to drill for oil or minerals.
- Seismologists are scientists who study earthquakes. A seismograph is an instrument that records vibrations in the ground. Seismograph stations around the world record the earth's vibrations. The epicenter of an earthquake can be located by computing the distance from each seismograph that recorded the quake.
- Seismologists monitor the San Andreas Fault daily. Creepmeters placed along the fault line record slight movements of the fault. Lasers show movements of the earth's crust. Magnetometers show changes in the magnetic properties of rocks. Seismographs record tremors that occur. All this information helps scientist predict any big quake that may occur.
- The San Andreas Fault enters California at Cape Mendocino. That is about 100 miles north of San Francisco. The fault runs south to the Gulf of California. Its full length is about 650 miles. If you saw it from an airplane, it would look like a big trough or valley. Much building has been done on or near the fault.
- Seismographs are very sensitive instruments. They can record tiny tremors thousands of miles away. Earthquake shocks show up as wiggly lines traced on a turning drum. A larger shock registers a larger wiggle.

  Seismograph stations are set up all over the world. Measurements obtained by the seismographs can be used to locate the focus and epicenter of quakes.

What Causes Earthquakes?

An earthquake is a violent shaking of the ground. Most earthquakes last less than a minute. No other kind of natural disaster causes such a large amount of damage in such a short time.

Millions of earthquakes take place around the world each year. Fortunately, only a few of these are strong enough to cause serious damage and loss of life.

Scientists believe that earthquakes begin deep inside the earth, in the middle layer called the mantle. As we have learned, the earth's crust is split into huge plates. The plates float about slowly on the molten rock of the mantle. As the plates move, they sometimes collide. These collisions are the main cause of earthquakes.

When two plates collide, each tries to push past the other. Tremendous pressure builds up in the mantle. When the stress becomes too great, something must give way. When it does, it does so violently. The site of this movement beneath the earth's crust is called the

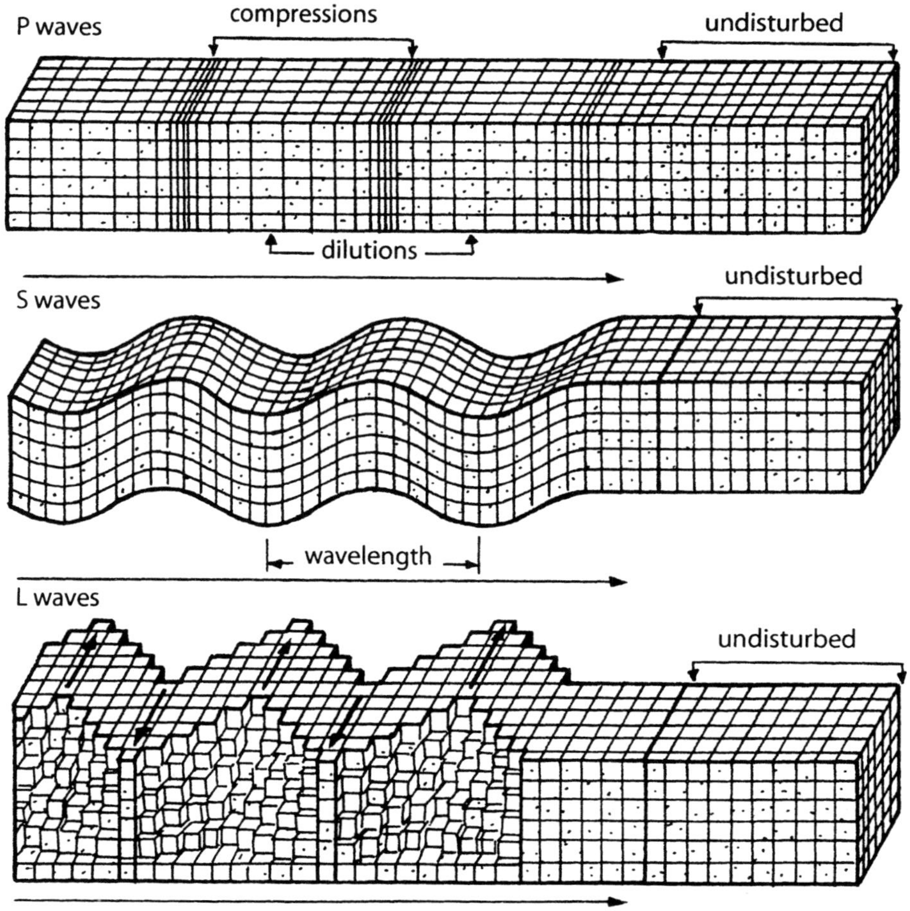

(continued)

What Causes Earthquakes? *(continued)*

earthquake's focus. The point on the earth's crust directly above the focus is called the epicenter of the quake.

Shock waves called seismic waves travel away from the epicenter in all directions very quickly. There are three main kinds of waves.

- **Primary waves** (P waves) travel the fastest. They move about 5 miles per second. They travel under the earth's surface. They cause rock particles to move forward and backward in the same direction as the wave is traveling, much like a steel coil expands and contracts.
- **Secondary waves** (S waves) are also called shear waves. They travel under the earth. S waves shake rock particles back and forth at right angles to the direction the wave is traveling. They are slower than P waves. They travel at about 3 miles per second.
- **Surface waves** (L waves) travel from the focus of the quake directly to the epicenter. Then they move along the surface of the earth, much like waves travel on the ocean. With each L wave, the earth's surface moves up and down like a wave. Thus, L waves cause most of the damage since they move the earth's surface up and down.

Earthquake shocks may last for less than a second in a small quake or for several minutes in a large one. Weaker shocks, called aftershocks, may come after a quake for days, weeks, or longer. These aftershocks are caused by the plates' continuing readjustments.

Where Do Most Earthquakes Occur?

Four out of five quakes take place in an area circling the Pacific Ocean. This area is called the Ring of Fire. It contains most of the world's active volcanoes.

A second earthquake zone runs from Southeast Asia through the Mediterranean Sea and out into the Atlantic Ocean.

These two areas have nearly all of the world's earthquakes. This is because they are areas where major plates are colliding. However, earthquakes do occur in other places.

The United States has its share of earthquakes. About 700 quakes rock this country each year. Half of these are in southern California. The only places that have never had an earthquake are the southern parts of Texas, Alabama, and Florida.

Earthquake Faults

Faults are cracks in the earth's surface where two plates meet. Most lie beneath the surface. But some can be seen as long cracks in the ground.

The most famous fault is the San Andreas Fault. It runs parallel to the coast of California for 700 miles. Here, the plate on the ocean side is grinding against the eastern-side plate on its way north at a rate of 2 inches a year. Scientists say that one day (in 10 million years), Los Angeles will be sitting alongside San Francisco.

Over 20 million people live along the San Andreas Fault. Dozens of quakes occur along the fault each year. Scientists believe a major quake may strike there in the near future.

What Are the Effects of Earthquakes?

Shocks and Aftershocks

Earthquake shocks can damage or destroy buildings, roads, bridges, and just about anything else they shake. Trees may be uprooted. Water may gush from broken dams. Aftershocks can shake down structures already weakened by the quake.

Landslides and Avalanches

In hilly or mountainous areas, the force of the quake may send loosened rocks or snow downhill. This can be very destructive. For example in Peru in 1971, a quake shook a large mass of ice and rock loose from a glacier. Some 50,000 people were crushed to death.

Quake-Related Fires

If a quake strikes a city, the resulting fires may cause more damage than the shaking. Gas lines may break and cause fires. Meanwhile, the quake may have broken water mains. Thus most firefighting equipment becomes useless. In 1923, an earthquake occurred near Yokohama and Tokyo, Japan. Yokohama, a city of 750,000, burned to the ground. Half of Tokyo was destroyed. About 95 percent of that damage was from fires. This catastrophic quake killed 150,000 people.

Tsunamis

Tsunamis are often called tidal waves. However, this term is incorrect because tsunamis have nothing to do with tides. Actually, these killer waves are triggered by undersea earthquakes. When an earthquake strikes under the ocean, it may send tsunamis in all directions. What happens is that the quake pushes the ocean floor upward toward the surface. An enormous mass of water is then displaced. This water travels across the ocean at incredible speeds—up to 500 miles per hour in the open ocean.

If you were to see a tsunami wave in the open ocean, it would not look very high. That is because most of the wave is under the surface of the water. The tsunami wave may actually be 4000 meters deep.

When this mass of water approaches the shore, the ocean floor slopes sharply upward. The moving water is forced upward as well. Thus, the wave becomes very high. For example, in 1737 a 210-foot tsunami hit the coast of Siberia. Most tsunamis occur in the Pacific and Indian oceans.

Changes in the Landscape

Quakes may cause big changes in the surface of the land. The ground on one side of a fault may be moved up or down. In the San Francisco quake of 1906, land on the west side of the San Andreas Fault was moved 20 feet northward.

How Are Earthquakes Measured?

Two different scales are used to measure the strength of an earthquake: the Richter scale and the Mercalli scale. With the Richter scale, scientific instruments called seismographs measure the magnitude or the amount of energy released by a quake. With the Mercalli scale, the amount of damage done by a quake is a subjective measure.

The Richter scale was developed by American geologist Charles Richter in 1935. Each whole number on the scale means that the earthquake causes 10 times as much ground motion and releases 32 times as much energy as the number below it. For example, a quake with a magnitude of 5 has ground waves 10 times as large as a quake with a magnitude of 4 and 100 times as large as a quake with a magnitude of 3.

	Richter Scale	**Mercalli Scale**
2.5	Nothing felt	I. Not felt II. Hanging lamps may sway
3.5	Felt by many people	III. Felt by some people indoors; many do not recognize it as an earthquake IV. Felt indoors by many; felt outdoors by a few V. Felt by most; many woken up
4.5	Some damage	VI. Felt by all; damage slight VII. Little damage in well-built buildings; much more in poorly built structures
6.0	Destructive earthquake	VIII. Damage slight in earthquake-proof buildings; collapse of chimneys, walls, columns and statues IX. Large cracks in ground; even well-built buildings off foundations
7.0	Major earthquake	X. Most structures destroyed; rail lines cracked; landslides on hills
8.0	Great earthquake	XI. Few buildings left standing; large cracks seen XII. Total damage; waves seen on the ground

Name ______________________________

Date ______________________________

EARTHQUAKES
REVIEW 2

Get the Facts

1. What do scientists believe causes earthquakes?

2. What is the epicenter of an earthquake?

3. What type of earthquake wave causes the greatest amount of damage?

4. Where do four out of five earthquakes take place?

5. What is a fault?

6. What is the famous fault in California?

7. What is the main difference between the Richter scale and the Mercalli scale?

8. What number on the Richter scale means a major earthquake?

9. Name four destructive effects of earthquakes.

What Do You Think?

10. An earthquake measuring 8.0 on the Richter scale is an extremely strong quake. But not all quakes with the same rating on the scale cause the same amount of damage or loss of life. Why do you think this is so?

Teacher's Guide for pp. 35–38

This teacher material covers the following reading selections:

Can Earthquakes Be Predicted?
Can Earthquakes Be Controlled?
How Can You Be Prepared for Earthquakes?

Vocabulary (See page 42 for definitions.)

conductor	radon
elevation	tremor

Answer Key (Review 3, page 38)

1. (any four) changes in the elevation or tilt of the land along a fault; changes in water level or water temperature in deep wells in the area; large amounts of radon in water in deep wells; surface rocks become better conductors of electricity; small earth tremors; water seeping up from the ground; changes in animal behavior
2. It provides funds for scientists to do research on earthquake prediction.
3. Pacific Tsunami Warning Center in Hawaii
4. They build on rock rather than clay, sand, or gravel. They build with reinforced concrete or with a steel frame.
5. Put an X by f, g, h.
6. Answers will vary.

Additional Information on Predicting, Controlling, and Being Prepared for Earthquakes

- Skyscrapers said to be earthquake-proof have been built in both Los Angeles and San Francisco. The First Interstate World Center in Los Angeles is 1017 feet tall. The Transamerica Pyramid in San Francisco is 853 feet tall.
- The U.S. Geological Survey's National Earthquake Information Service is located in Golden, Colorado. This service collects earthquake information from 3000 seismological stations in 125 countries. One of its goals for the future is the prediction of earthquakes.
- Scientists are working on ways to reduce the power of earthquakes. One idea is to lubricate the plates with water so that they slip past each other more smoothly. Water might be pumped to the plates from wells drilled deep into the faults. If this lessened the buildup of pressure along the fault, earthquakes would be less severe.

Can Earthquakes Be Predicted?

For a number of years, scientists have been trying to identify signs that can be used to predict when earthquakes will occur. Signs are related to increasing stress in the earth's crust. Some of these signs are:

- changes in the elevation or tilt of the land
- changes in water level or water temperature in deep wells
- large amounts of radon in water in deep wells
- surface rocks conducting more electricity
- small earth tremors
- water seeping up from the ground
- changes in animal behavior

Animals seem to be able to sense a quake before it occurs. Strange behavior has been observed in many species of wild and domestic animals. Snakes and rats come out of their holes. Dogs bark all night or run away. Birds fly off in an agitated manner.

Scientists continue to study earthquake prediction. In 1977, Congress passed the National Earthquake Hazard Reduction Program Act. This provides funds to scientists so they can do research in this area.

The Pacific Tsunami Warning Center in Hawaii monitors all earthquakes. If a quake is likely to cause tsunamis, the center alerts all Pacific Rim countries.

Can Earthquakes Be Controlled?

We are not able to control earthquakes. But we can do things to lessen the damage they cause.

Most deaths from earthquakes are caused by the collapse of buildings. Better construction methods can make buildings more earthquake proof. In areas like California and Japan, new houses are built to be earthquake resistant.

Architects know how to make buildings safer. They build on solid rock rather than on clay, sand, or gravel. For tall buildings, they use reinforced concrete or a steel frame. They tie together as one unit all parts of the structure. They join walls, floors, joints, and supports. Then when an earthquake hits, the structure vibrates as a unit, rather than shaking apart.

How Can You Be Prepared for Earthquakes?

Even if you don't live in earthquake country (California), you should be concerned about earthquakes. They can happen in most states . . . without warning. Knowing what to do can save your life.

Most deaths and injuries in an earthquake are not caused by the shaking earth. They are caused by the collapse of buildings, falling objects, and fires from broken gas or electrical lines. A few precautions taken ahead of time can reduce these risks.

Check Your Home Before an Earthquake Strikes

Check your home ahead of time for possible dangers. Take care of any problems you find.

Find out how to turn off electricity, gas, and water to your home. Your local utility companies can give you this information. Repair any defective electrical wiring or leaky gas connections. Bolt down gas appliances and water heaters.

Make sure shelves are securely fastened to walls. Heavy items should be placed on lower shelves. Breakable items should be stored in closed cabinets or on low shelves.

Make sure overhead fixtures like lights, fans, or plants are securely connected.

Repair any deep cracks in ceilings or foundations.

Prepare Your Family

If an earthquake occurs, you may find it is impossible to move from the room you are in. So, it is important to know safe spots as well as areas to stay away from in each room.

Walk through your house. Try to find a safe spot in each room. Some of the best places are under pieces of furniture. Look for spots under a desk or table, in a doorway, or against an inside wall. Stay away from windows, mirrors, fireplaces, bookcases, wall units, and objects that hang overhead.

Collect some basic emergency supplies: a flashlight and extra batteries, a battery-operated radio, a supply of drinking water in case the water supply is contaminated, canned foods that can be eaten without cooking, a fire extinguisher, and a first aid kit. Keep these supplies where they can be easily found.

What to Do During an Earthquake

Most injuries during earthquakes take place when people are going in or out of a building. So, if you are inside when the quake begins, stay

(continued)

How Can You Be Prepared for Earthquakes? *(continued)*

inside. Take cover following the plan you have already made. If you are using the stove, turn it off. Put out any fires.

If you are in a tall building, find a safe spot on the floor you are on. Stay away from windows and outside walls. Do not try to use the elevator. Avoid the stairs, which may be broken or jammed with people.

If you are outside, stay outside. Move away from buildings, trees, and electrical wires. The greatest danger from falling debris is close to buildings. If you are in your car, pull over and stop. Stay in your car. Don't park under bridges or wires.

What to Do After an Earthquake

Check for injuries. Perform any needed first aid. Turn on your radio or television to get emergency bulletins.

Check your gas, water, and electrical lines. If you smell gas, open the windows and shut off the main gas valve. Then leave the building. Stay out until it has been checked by the gas company. Don't use candles, matches, or other open flames until you are sure there has been no gas leak.

If electrical wiring is damaged, turn it off at the main meter box. Stay away from downed power lines.

If water pipes are broken, shut off the water supply. You can get emergency water from your hot-water heater.

Phone lines and roads need to be kept clear for emergency personnel. So stay off the telephone unless there has been a serious injury. And don't drive your car unless absolutely necessary.

Inspect your home for damage. Open doors to cupboards and closets with care. You could get injured by objects falling off the shelves. Look over any fireplaces carefully. If there are cracks, don't use the fireplace until it has been repaired.

Be prepared for aftershocks. These are usually smaller than the first quake. But they may be strong enough to cause more damage. If your home is unsafe, go to a community shelter.

Stay away from low-lying coastal areas until you hear that any danger of a tsunami has passed.

Name ______________________________

Date ______________________________

EARTHQUAKES
REVIEW 3

Get the Facts

1. Name four signs that may predict that an earthquake will occur.

2. What is the purpose of the National Earthquake Hazard Reduction Program Act?

3. What program has been set up to warn Pacific countries of possible tsunamis?

4. Name two ways architects design buildings to make them more earthquake resistant.

5. Put an X by each correct thing to do in the event of a quake.

During a Quake

_____ (a) If you're inside, run outside quickly.
_____ (b) If you're driving, take shelter under a bridge.
_____ (c) If you smell gas, light a match to look for the leak.
_____ (d) If you're in a tall building, run down the stairs.
_____ (e) Find a safe spot such as a doorway or outside wall.

After a Quake

_____ (f) If you smell gas, turn it off and open windows.
_____ (g) If wiring is damaged, turn off electricity at the meter.
_____ (h) Use your hot-water heater for emergency water.
_____ (i) Call up your friends right away.
_____ (j) Drive to the beach to look for tsunamis.

What Do You Think?

6. What do you think are the safest spots in your home? List where you would seek shelter in each room of your home in the event of an earthquake.

Earthquakes

Additional Information

Some Major Earthquakes

Date	Location	Deaths
1556	Shensi, China	830,000 die in the deadliest earthquake of all time
1703	Tokyo, Japan	200,000 killed
1737	Calcutta, India	300,000 killed
1755	Lisbon, Portugal	60,000 die as 50-foot tsunami washes away city
1811	New Madrid, Missouri	quakes open huge cracks in earth, 10 feet wide, 5 feet deep, 5 miles long
1906	San Francisco, California	America's most famous quake; city burns for four days; 600 killed
1908	Messina, Sicily	four-story-high tsunami drowns 100,000
1923	Tokyo and Yokohama, Japan	8.3 quake destroys all of Yokohama and half of Tokyo, as both cities burn for two days
1939	Chile	30,000 killed
1939	Turkey	100,000 killed
1960	Chile	8.4 quake causes 24-foot-high waves; 5000 killed
1962	Iran	12,000 killed
1963	Yugoslavia	1100 killed
1964	Alaska	8.5 quake shakes 1 million square miles
1968	Iran	12,000 killed
1972	Managua, Nicaragua	10,000 killed
1970	Peru	70,000 killed in avalanche from a glacier
1976	Guatemala City, Guatemala	23,000 killed
1976	Tangshan, China	250,000 killed
1978	Iran	25,000 killed
1985	Mexico City, Mexico	7000 killed
1988	Soviet Armenia	25,000 die; 500,000 homeless
1989	San Francisco, California	quake during World Series kills 62; great property damage
1990	Iran	50,000 killed; 400,000 homeless
1992	Indonesia	2500 killed on island of Flores
1993	Japan	300 dead
1995	Kobe, Japan	3000 killed

(continued)

Additional Information *(continued)*

Earthquakes in Ancient Times

In ancient times, people believed that the earth was the center of the universe. The sun, moon, and stars revolved around the earth. The earth was held up by some sort of support, usually an animal or a god. If that support moved, an earthquake resulted.

In Japan, it was believed that a giant spider held up the earth. When the spider moved, an earthquake occurred. Later, the Japanese believed it was a giant catfish that held up the earth.

In Mongolia, the belief was that a giant hog shook the earth when it scratched itself on a palm tree. In India, it was a giant mole. In South America, it was a whale. For North American Indians, it was a giant tortoise. In the East Indies, it was a water buffalo.

In Siberia people said earthquakes were caused by the dogs of the god Tuil, rolling about to rid themselves of fleas.

Another legend told of the earth being circled by snakes. If one snake began chasing its tail, the earth would shake.

The ancient Greeks thought the earth rested on the shoulders of Atlas. If Atlas shrugged his shoulders, he caused earthquakes.

Aristotle wrote about some of the ideas for the causes of earthquakes in his day. One theory said there were great winds blowing under the surface of the earth. If the winds became bottled up anywhere, they would erupt to the surface.

Others of Aristotle's time postulated the existence of huge underground caverns. When these caverns collapsed, earthquakes resulted.

The Ancient Mystery of Atlantis

Ancient writers told the story of the lost continent of Atlantis. Plato wrote of Atlantis as an island that sank under the sea. Ancient Egyptian lore tells of an island off the coast of Greece which was swallowed up by the sea overnight.

For years people wondered about the mysterious Atlantis. Modern exploration and scientific thought seem to have solved the mystery. Atlantis may have been an island off the coast of Greece. Archeologists have discovered ruins of a fabulous Minoan civilization on the island of Thera. This island and its neighbors were buried by the sea in the aftermath of an earthquake followed by a volcanic eruption.

The Earth 200 Million Years Ago

Once, all the present continents of the earth were part of one large landmass called Pangaea. The land was surrounded by one large ocean. Then, about 200 million years ago, the landmass broke into pieces. Eventually, the continents reached the positions they are in today. This theory, proposed by Alfred Wegener in 1912, has been supported by fossil and geological evidence (see page 7).

The First Seismograph

In A.D. 132, a Chinese astronomer named Chang Heng developed an earthquake weathercock. This bronze instrument was shaped like a jar. It was 8 feet tall and 3 feet wide. It had eight dragon heads around the outside. Each head had a ball in its mouth. When an earthquake struck, the weathercock vibrated. When the quake was strong enough, a bronze ball fell from one dragon's mouth into a frog's mouth below. This showed not only that an earthquake had occurred but also which direction it had come from. The instrument was used for the next 1600 years.

Activities for Students

1. Write to the Geological Survey Office at your state capital. Ask where the faults occur in your state. Ask if any earthquakes have occurred in your state. If so, ask where they have occurred. Mark them on a map of your state.
2. Report on one of the major earthquakes that have occurred in the world. Use the list on page 39.
3. Watch newspapers for stories about earthquakes around the world. Locate where they occur on a map. Find out if they are on the edge of one of the crust's plates.
4. Read more about the San Andreas Fault. Have there been any recent changes that would suggest that major shifts are about to occur? How are changes on the fault monitored?
5. Find out what is being done in California to predict future earthquakes.
6. Research the link between volcanoes and earthquakes.
7. Find out more about the United States Geological Survey.
8. Write to one of the following for more information about earthquake safety:

   U.S. Geological Survey
   Public Inquiries Office
   302 National Center
   Reston, Virginia 22092

   Federal Emergency Management Agency
   Natural Hazards Division, ONTH-SLPS
   500 C Street, S.W.
   Washington, D.C. 20472

   Your local American Red Cross Chapter
9. Find out more about early earthquake-detecting instruments.
   - Paul Gabriel Hautefeuille's design of the early 1700's that improved on Chang Heng's weathercock
   - Luigi Palmieri's 1853 invention of a clocklike instrument that stopped when a shock hit and set in motion a recording drum to measure the shock's length
   - John Milne's 1897 invention of the seismograph
10. Read more about Atlantis, the lost continent.

Earthquake Vocabulary

aftershocks	smaller quakes following an earthquake
avalanche	rapid downhill movement of rocks or snow
conductor	material through which electricity passes
earthquake	sudden movement and shaking of the earth's crust
elevation	altitude of a place
epicenter	point on the earth's surface directly above the focus of an earthquake
fault	crack in the earth's crust where rocks move against each other
focus	place in the mantle where the earthquake has its center
L wave	surface wave
mantle	middle layer of the earth
Mercalli scale	scale for measuring the effects of an earthquake in a given place
P wave	underground earthquake wave
plate	one of the large pieces of the earth's crust
radon	gas formed by the disintegration of radium
Richter scale	scale that measures the magnitude of quakes
Ring of Fire	earthquake and volcano belt that runs around the Pacific Ocean
S wave	secondary waves under the earth that vibrate from side to side
seismic wave	wave of energy produced by an earthquake
seismograph	instrument that records the intensity, time, duration, and direction of a quake
tremor	slight shaking or vibration
tsunami	huge ocean wave set in motion by an earthquake

When the Wind Blows

Teacher's Guide for pp. 45–46

This teacher material covers the following reading selection:
The Xenia, Ohio Tornado

Vocabulary (See page 61 for definitions.)

F5 tornado
National Severe Storms Forecast Center

Answer Key (Review 1, p. 46)

1. A cold front was moving on a collision course with warm, moist air from the Gulf of Mexico. Such conditions are perfect for the development of storms.
2. alerted weather stations in 16 states; sent weather balloons aloft; put weather satellites on "severe weather mode," which enabled them to take more photographs of the storm system
3. 148
4. winds up to 300 mph; cut a path 2000 to 3000 feet wide; hit for about three minutes
5. She got them to take cover in an appropriate area of the school.

Answers for the following will vary.

6. If there is a severe weather warning, you can change plans that might put you in a dangerous situation. Also, you can stay tuned to hear further developments in the weather.
7. The best places in a school would be an interior hallway with no windows, or a small windowless interior room like a closet. It is best if the hallway opens on the east or north. These sides usually have less wind. Tornadoes usually come out of the southwest and go toward the northeast. Thus, a hallway that opens toward the southwest could become a wind tunnel for the tornado to blow through. Avoid large rooms such as cafeterias, gyms, or auditoriums.

Additional Information on the April 1974 Tornadoes

- A large room such as an auditorium, cafeteria, or gymnasium is generally not a safe place to be in a tornado or hurricane. The large roof span makes it susceptible to collapse.
- Other towns that were nearly wiped out included Brandenburg, Kentucky, and Guin, Alabama.
- In all, 315 people were killed by the tornado system and 6200 were injured.
- The President declared 10 states disaster areas.
- Six of the tornadoes on April 3 were rated as F5 tornadoes. Three others were rated F4. Over 80 percent of the deaths from the tornado system were caused by the F4 and F5 storms.
- Nearly a fifth of the people hurt by the tornadoes were in mobile homes. These homes are one of the most dangerous places to be in a tornado.
- Scientists who studied these tornadoes later were able to disprove earlier theories that tornado winds go up to 500 mph. Actually, they rarely top 250 mph.

The Xenia, Ohio, Tornado

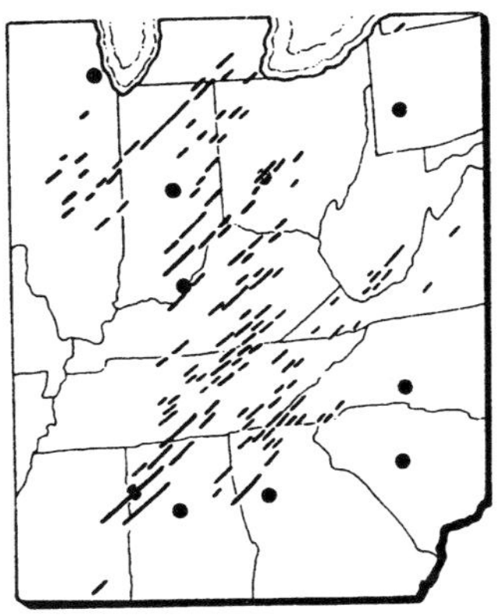

It was April 2, 1974. Meteorologists at the National Severe Storms Forecast Center were worried. They were monitoring a massive cold front. The front was moving into the Midwest. It was moving toward a head-on collision with warm, moist air from the Gulf of Mexico. Such conditions usually lead to severe thunderstorms and tornadoes.

As meteorologists studied the developing storm system, they knew that the potential was there for some severe storms. But they couldn't tell when and where the problems might develop.

Weather stations in 16 states were alerted. Additional weather balloons were sent aloft. Weather satellites went on severe-weather mode, enabling them to take additional photographs of the storm system.

On April 3, the storm system broke loose. All that afternoon and into the morning of the next day, one tornado after another swooped across a 13-state area. In all, 148 tornadoes were sighted in a two-day period.

Xenia is a small Ohio city of 27,000 people. Xenia was directly in the path of a tornado. And that tornado was an F5, the most powerful category of tornado, with winds over 261 mph. At 4:40 P.M. the tornado made a direct hit on Xenia.

For just over three minutes, the tornado ripped through the city, with winds close to 300 mph. It uprooted huge trees and tossed them about like matchsticks. It flattened schools, churches, houses, and offices. It left a path of distruction 2000 to 3000 feet wide.

The tornado killed 34 Xenia residents. It injured another 1600. It destroyed nearly 3000 homes. And it caused damage of $75 million.

The Xenia tornado could have been even worse. Hundreds of children might have died if the tornado had struck two hours earlier while school was in session. Three of the eleven schools in the town were completely destroyed. Two others were badly damaged.

As it was, the schools were not totally empty. At Xenia High School, the drama club was practicing in the school auditorium. Student Ruth Venuti was standing outside the school. She was waiting for a ride home. She saw a funnel-shaped cloud in the sky about two miles away. Suddenly, she realized it was a tornado. She ran to warn the other students. At first, the others thought she was joking. But when they looked outside and saw cars bouncing around in the parking lot, they knew her warning was real. The students and their teacher ran to the center hall of the school, crouched against the walls, and covered their heads.

Glass shattered. Dirt, wood, and sections of lockers flew down the hall. The noise was deafening. There was a momentary calm as the eye of the storm passed. Then the pandemonium began again. Finally it was over. No one had been seriously hurt. But when the students returned to the auditorium, they saw that the roof had collapsed onto the auditorium seats. And on the stage where they had been practicing was a schoolbus, turned upside down.

Name ______________________________

Date ______________________________

TORNADOES
REVIEW 1

Get the Facts

1. Why were meteorologists at the National Severe Storms Forecast Center concerned about the weather on April 2, 1974?

2. What three actions did the center take to better monitor the storm?

3. How many tornadoes were sighted on April 3 and 4?

4. Describe the tornado that struck Xenia, Ohio, on April 3.

5. How did Ruth Venuti's actions probably save the lives of fellow students at Xenia High School?

What Do You Think?

6. It is not usually possible to forecast the path or timing of a tornado. This makes the meteorologist's job difficult. What do you think is the value of generalized severe weather or tornado warnings?

7. Where would be the safest areas in your school if a tornado were to strike?

Teacher's Guide for pp. 49–53

This teacher material covers the following reading selections:

How Are Thunderstorms and Tornadoes Formed?
What Happens During a Tornado?
How Are Tornadoes Classified?
What Causes Lightning?

Vocabulary (See page 61 for definitions.)

counterclockwise
cumulonimbus clouds
cumulus clouds
downdraft
Fujita-Pearson Tornado Intensity Scale
funnel cloud
lightning
mesocyclone
supercell
thunderstorm
tornado
updraft

Answer Key (Review 2, p. 53)

1. in the midwestern United States
2. April, May, or June
3. cumulonimbus clouds
4. High winds blow the cooler air in an easterly direction while lower winds blow the warmer air toward the north. This may cause a column of air to spin.
5. counterclockwise
6. (any three) very dark or greenish sky; hail may have fallen; momentary calm; a wall cloud; a thunderstorm has just stopped
7. lightning, hail, flash flooding
8. by their wind speed and amount of damage done
9. F4 and F5
10. 20 to 50 mph

Answers for the following will vary.

11. Tornadoes can do strange things that later seem almost amusing. They have blown the feathers off a chicken. They have picked up persons and deposited them unhurt miles away. And they have moved a house to a nearby lot. But more often, tornadoes are extremely destructive of lives and property.
12. All structures should be built to withstand strong winds. Public shelters should be available underground. Large office buildings in cities should have shelter areas. Homes should have a basement, an outside storm cellar, or a windowless room or closet that can serve as a shelter.

Additional Information on Thunderstorms and Tornadoes

- Tornadoes are also called twisters or cyclones.
- A wall cloud is a veil of unusually heavy rain that comes out of the leading edge of the cumulonimbus cloud. A wall cloud can be one of the visual signs of an approaching tornado.
- Thunderstorms can be hazardous to aircraft. In general, the Federal Aviation Administration directs planes around storms. However, planes may have to pass through a storm when taking off or landing. Thunderstorms may produce severe downdrafts and turbulence that can cause an airplane to lose altitude rapidly. Wind shear is large variation of wind velocity in a small distance. This is another cause of major airline crashes.
- A partial vacuum exists within the tornado. Scientists are not sure how low the pressure actually is within the tornado. It may be as low as 18.8 inches of mercury. In any case, tornadoes are able to suck up large objects with ease.

Tornadoes

- Most tornadoes come from the southwest and move toward the northeast. The average length of their path is 4 miles. But some have traveled hundreds of miles. The width of their path is about 300 to 400 yards on the average. But it may be as wide as a mile.
- Texas, Oklahoma, and Kansas have the largest number of tornadoes. Mississippi has the most deaths from tornadoes. This fact may be due to the number of Mississippi residents who live in mobile homes.
- The United States usually has 500 to 600 tornadoes each year. Some years, the number is much higher. In 1967, more than 900 were reported. In 1973, 1100 were reported. But 1974 had the most ever (see pages 45–46). Some of this increase may simply be due to the improvement of instruments for detecting tornadoes.
- The Skywarn system uses trained tornado observers who spot tornadoes and report them to the National Weather Service.
- Toto is the National Weather Service's Totable Tornado Observatory. It is named after Dorothy's dog in the *Wizard of* Oz. Toto is a package of weather-recording instruments. The instruments can be taken to a place where a tornado is expected. It measures temperature, atmospheric pressure, and wind speed and direction.
- A few severe thunderstorms are very large and strong. They cover an area of more than 100 square miles. They are called supercell thunderstorms. Supercells can last more than three hours.
- Hail is formed when updrafts pull rain high into the air. There, the air is cold enough to freeze the rain.
- Waterspouts are tornadoes that occur over water. The spiraling wind picks up water and swirls it around.
- Tornadoes accompany hurricanes when clouds break away from the hurricane and spawn thunderstorms.
- Tornadoes are of such short duration that scientists have problems getting accurate data about them.
- Tornado Alley is the name given the area of Texas, Oklahoma, and Kansas that is hit by so many tornadoes.
- Tornadoes can easily destroy mobile homes and small buildings. In regions where tornadoes are common, structures should have basements or storm shelters and should be able to withstand high winds.
- Hail has the potential to ruin crops. So, for centuries, people have tried to reduce its damaging effects. In the sixteenth century, people rang church bells and fired artillery to ward off severe thunderstorms. Similar attempts have continued in the twentieth century, with little effect. Modern efforts include hailstorm seeding. Seeding is based on the idea that a cloud contains a limited amount of moisture. If that moisture falls as small hailstones rather than large ones, damage would be less. And some of the smaller stones might even melt on the way to earth. Results of such experiments have been mixed.
- Lightning causes forest fires. So people have experimented with ways to decrease lightning. One way is to seed storms with silver-iodide nuclei. Other studies for reducing lightning are ongoing.

How Are Thunderstorms and Tornadoes Formed?

Thunderstorms occur most often in spring and summer. They are the most important source of water for crops of the American Midwest. This area of is known as the breadbasket of the world.

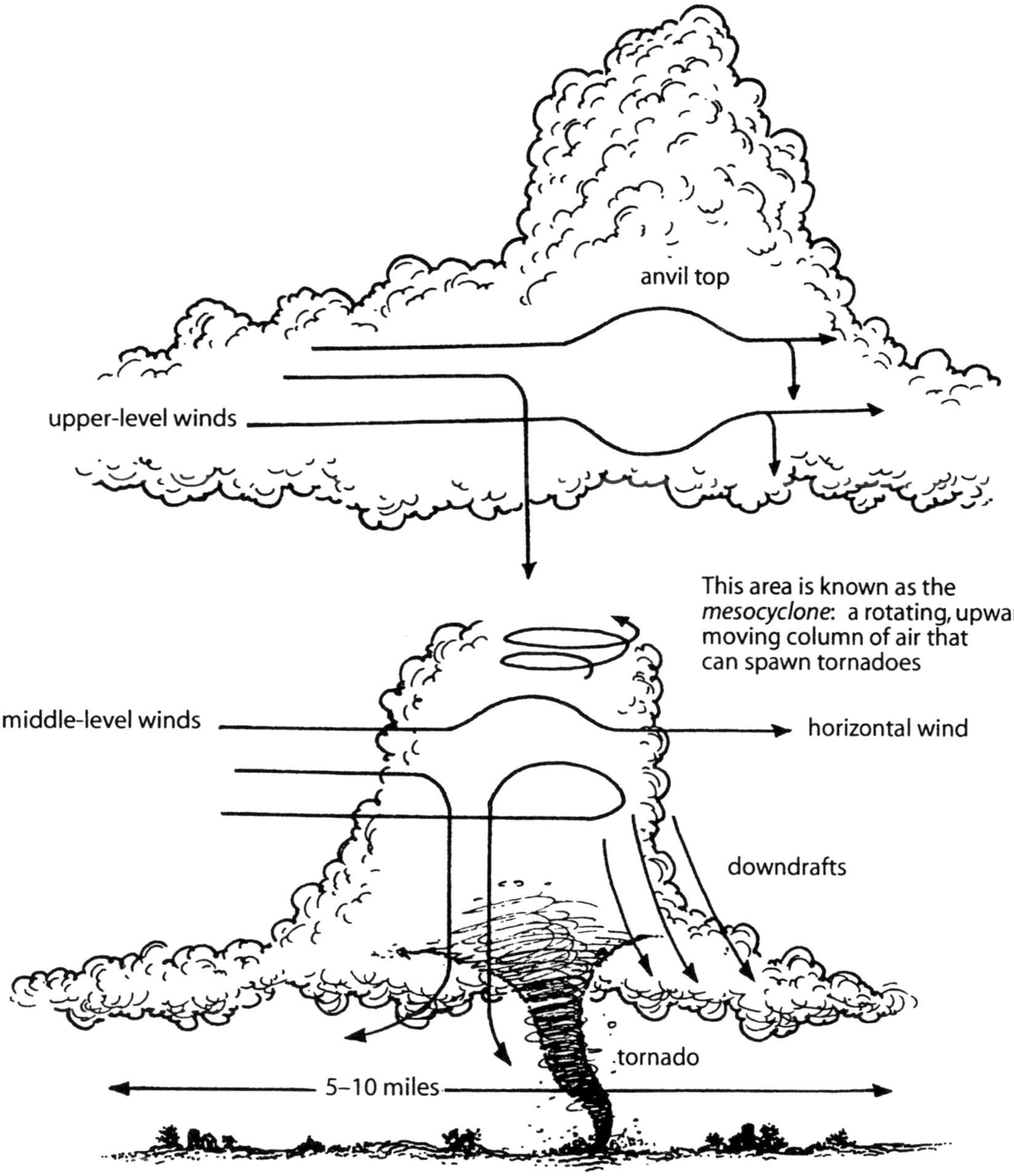

A Supercell: King of Thunderstorms

(continued)

How Are Thunderstorms and Tornadoes Formed? *(continued)*

In this area of the United States, cold air from the Arctic meets warm air from the tropics. Normally, the warm air rises and the cold air sinks. The warm air carries a great deal of water vapor from the Gulf of Mexico. As it rises, it cools. The water vapor condenses into droplets. These droplets form fluffy white cumulus clouds.

As the clouds pick up more water, they become larger and darker. They are now called cumulonimbus clouds. From these clouds, thunderstorms are formed. They may produce thunder, lightning, wind, heavy rain, and sometimes hail.

During a storm, columns of warm air called updrafts rise. This warm air adds more moisture to the storm system. Columns of cold air called downdrafts bring rain to the earth. The thunderstorm continues until all the moisture has fallen from the clouds.

Most thunderstorms last less than an hour. But some of the larger ones can last several hours. These are called supercell thunderstorms. Supercells produce violent weather such as tornadoes and hailstorms.

A supercell has strong, persistent updrafts and downdrafts. High above the earth, winds blow the cool air from west to east. Nearer the ground, winds blow in a northerly direction. In a supercell, these winds blowing in different directions may cause an updraft to begin spinning. Then it is called a mesocyclone, or funnel cloud. As the mesocyclone continues to grow, it spins faster and faster. It becomes longer and narrower. It stretches down toward earth. It has become a tornado.

Most tornadoes are funnel-shaped—wider at the top and narrower at the bottom. Some are shaped like a cylinder. Others look like a long rope. Usually they spin in a counterclockwise direction.

About 90 percent of all tornadoes in the world occur in the midwestern United States. Oklahoma, Texas, and Kansas have the most. Tornadoes are most numerous and severe during April, May, and June. A second weaker tornado season takes place in October and November.

What Happens During a Tornado?

Warning Signs

Tornadoes often strike just after a thunderstorm has passed. The sky is usually very dark. It may even appear greenish. Often, large hail has fallen. There may be a wall cloud hiding tornadoes behind it. Things appear calm, as if the storm is over. But the worst is about to come.

A loud roar begins. Some say it sounds like a squadron of jet planes. Others say it sounds like a freight train overhead.

What Tornadoes Do

The danger from a tornado comes from a dual effect: very strong outer winds and extremely low inner pressure. Pressure differences inside and outside a building can cause the building to literally explode. At the same time, the building is being hit by furious winds that can pick up and carry anything in their path.

Very low inner pressure makes the tornado act like a giant vacuum cleaner. It sucks up everything in its path. In *The Wizard of* Oz, by Frank Baum, a tornado sucked up Dorothy and her dog, Toto, and carried them all away to the land of Oz. In real life, tornadoes have picked up almost everything you can imagine and then set the items down often miles away. In Mississippi in 1991, a tornado picked a baby out of her crib and left her hanging by her diaper outside in a tree. The baby was unharmed.

Tornadoes travel in an erratic fashion. They may hit the ground once and disappear. They may strike, rise into the air, then strike again miles away. Sometimes they move in a straight line. Sometimes they hop about. Generally,

they travel at a speed of 20 to 50 mph. But their winds blow at 100 to 250 mph or more.

Tornadoes are often accompanied by other destructive forces. Lightning can cause damage, fire, or death. Hailstones as big as grapefruit can ruin property and crops. Flash flooding from torrential rains is the natural disaster that has caused the most deaths per year in the United States.

Tornadoes tend to occur in groups. If there is one, there are probably more. In just two days in April 1974, 148 tornadoes were seen over the eastern half of the United States. They killed 315 people and caused $600 million in property damage. They were the largest outbreak of tornadoes to occur in the United States in recorded history. To find out more about this record-setting outbreak, read pages 45–46.

Tornadoes

How Are Tornadoes Classified?

The Fujita-Pearson Tornado Intensity Scale is used to classify tornadoes. It ranks them by wind speed from F0 to F5. Luckily, most tornadoes are rated F0 or F1. Only 2 percent of tornadoes are ranked F4 or F5. But all kinds of tornadoes can damage property and kill people.

Rank	Wind Speed (mph)	Damage
F0	40–72	Slight
F1	73–112	Moderate
F2	113–157	Considerable
F3	158–206	Severe
F4	207–260	Devastating
F5	over 261	Incredible

About 70 percent of all tornadoes are F0 or F1. These tornadoes are of short duration. They last from 1 to 10 minutes. They account for only 4 percent of all tornado deaths.

About 28 percent of all tornadoes are F2 or F3. They usually last 20 minutes or longer. They account for nearly one third of tornado deaths.

Only 2 percent of all tornadoes are classified F4 or F5. They can last up to an hour or more. They account for two thirds of tornado deaths.

What Causes Lightning?

Lightning often accompanies thunderstorms. It is one natural disaster against which we know how to protect ourselves. Yet more people in the United States are killed by lightning each year than are killed by hurricanes and tornadoes put together.

In the year 1752, Benjamin Franklin proved that lightning was electricity. His proof was his famous kite experiment. He was lucky not to be electrocuted.

Scientists still do not understand how a thundercloud becomes electrostatically charged. But they do know that the upper and lower parts of a cloud generally have a positive charge, while the middle parts have a negative charge. When enough charge has accumulated, giant electrical sparks (lightning) flash between one area of the cloud and another. Most of these sparks begin in the center of the cloud and move toward the ground.

Teacher's Guide for pp. 49–53

This teacher material covers the following reading selections:

How Are Thunderstorms and Tornadoes Formed?
What Happens During a Tornado?
How Are Tornadoes Classified?
What Causes Lightning?

Vocabulary (See page 61 for definitions.)

counterclockwise
cumulonimbus clouds
cumulus clouds
downdraft
Fujita-Pearson Tornado Intensity Scale
funnel cloud
lightning
mesocyclone
supercell
thunderstorm
tornado
updraft

Answer Key (Review 2, p. 53)

1. in the midwestern United States
2. April, May, or June
3. cumulonimbus clouds
4. High winds blow the cooler air in an easterly direction while lower winds blow the warmer air toward the north. This may cause a column of air to spin.
5. counterclockwise
6. (any three) very dark or greenish sky; hail may have fallen; momentary calm; a wall cloud; a thunderstorm has just stopped
7. lightning, hail, flash flooding
8. by their wind speed and amount of damage done
9. F4 and F5
10. 20 to 50 mph

Answers for the following will vary.

11. Tornadoes can do strange things that later seem almost amusing. They have blown the feathers off a chicken. They have picked up persons and deposited them unhurt miles away. And they have moved a house to a nearby lot. But more often, tornadoes are extremely destructive of lives and property.
12. All structures should be built to withstand strong winds. Public shelters should be available underground. Large office buildings in cities should have shelter areas. Homes should have a basement, an outside storm cellar, or a windowless room or closet that can serve as a shelter.

Additional Information on Thunderstorms and Tornadoes

- Tornadoes are also called twisters or cyclones.
- A wall cloud is a veil of unusually heavy rain that comes out of the leading edge of the cumulonimbus cloud. A wall cloud can be one of the visual signs of an approaching tornado.
- Thunderstorms can be hazardous to aircraft. In general, the Federal Aviation Administration directs planes around storms. However, planes may have to pass through a storm when taking off or landing. Thunderstorms may produce severe downdrafts and turbulence that can cause an airplane to lose altitude rapidly. Wind shear is large variation of wind velocity in a small distance. This is another cause of major airline crashes.
- A partial vacuum exists within the tornado. Scientists are not sure how low the pressure actually is within the tornado. It may be as low as 18.8 inches of mercury. In any case, tornadoes are able to suck up large objects with ease.

- Many airplanes have been struck by lightning. Generally the damage has been slight. Danger comes if the plane's fuel catches fire. This has occurred only twice, as far as is known: in Milan, Italy, in 1959, and in Maryland in 1963.
- Space rockets also have been struck by lightning. In 1987, lightning struck a rocket. The mission was scuttled. In 1983, it almost struck the space shuttle. In 1969, it struck *Apollo 12* twice right after liftoff. But the mission went on.
- A "bolt from the blue" is a lightning bolt that strikes out of a clear, blue sky. Sometimes lightning occurs in a ball form. Scientists are not sure how this occurs.
- A lightning rod can attract lightning away from a house or building and lead it harmlessly to the ground. Why this works is not really understood. But it does work.
- Weather information can be obtained via radio or television by tuning in to:

  NOAA Radio
  The Weather Channel (cable TV)
  AM Weather (PBS)
  Your local radio and television stations

How Can You Get Ready for a Tornado?

Being Prepared for a Tornado

If a tornado strikes, you will have little or no warning. Knowing what to do ahead of time could save your life.

First, decide where in your home is the safest spot in case of a tornado. If you have a basement, that is your safest spot. Since most tornadoes approach from the southwest, the southwest corner of your basement is the safest spot of all. Any debris would be blown away from this corner. Just be sure there are no windows or glass doors near this spot.

If you have no basement, the next best choice is an interior room with no windows. This could be a hallway, bathroom, or closet on the ground floor. Get under something sturdy.

If you are in a multistory building, you may not have time to get to the basement or ground floor. In this case, go to a hallway in the center of the building, avoiding windows on the way.

If you are in a mobile home, leave it immediately. Find shelter elsewhere, or lie down outside in a low area. Cover the back of your head and neck with your hands. Never remain in a mobile home during a tornado or hurricane.

Put together a disaster supply kit. It should contain a flashlight, extra batteries, canned food, a can opener, bottled water, and a first aid kit. It is a good idea to keep instructions on how to turn off your home's utilities in this kit.

Review your tornado-preparation plan periodically. Make sure everyone in the family, especially children, remembers what to do in case of a tornado.

What to Do During a Tornado

Listen to your local radio or TV stations for weather bulletins. If a tornado watch is issued, a tornado is possible in your area. Stay tuned for more information.

If a tornado warning is issued, go to the safe place you have decided on. If you are outside, take shelter. If this is not possible, lie flat in a ditch or other low-lying area. If you are in your car or a mobile home, get out and take shelter. Don't try to outrun a tornado in your car.

After the Tornado

Listen to the radio for instructions or safety information. Watch out for fallen power lines. Stay out of areas that were damaged by the tornado unless you have a good reason to be there.

Tornado Warning Systems

The National Severe Storms Forecast Center is located in Kansas City, Missouri. Meteorologists there monitor weather conditions. They look for areas where tornadoes are likely to develop. Tornado watches are issued to the local weather stations. Local meteorologists search for tornadoes using Doppler radar as well as trained observers. If tornadoes are seen, tornado warnings are broadcast on radio and TV. Tornado sirens may be used to sound a warning.

Doppler radar is able to identify tornadoes within a thunderstorm early while they are still forming. This technology allows meteorologists to give people more advance warning of a tornado's approach. Currently, a network of Doppler radar is being installed across the United States.

What Can You Do in Severe Thunderstorms?

Many people are killed by lightning each year. If you hear thunder, you are close enough to a storm to be in danger from lightning.

In a thunderstorm, take shelter in a building or car. Avoid using your telephone or electrical appliances. Unplug appliances and computers. Electricity may travel in telephone or power lines. Power surges may ruin expensive equipment.

Avoid running water anywhere in the house. Do not take a bath or shower. Water is an excellent conductor of electricity.

If you are caught outside in a storm, take shelter. If you're in a boat or swimming in the water, get to shore and take cover. Find a low-lying open place, away from trees and metal objects. If you are in the woods, avoid tall trees and take shelter under a short tree.

If you are in an open area and can't get to shelter, crouch down. Place your head between your knees. Don't lie flat on the ground. That makes you a bigger target for the lightning to strike.

Persons who have been struck by lightning should be seen by a doctor as soon as possible. Call 911 or the emergency number for your area immediately. The victims can be touched safely. They carry no electrical charge. Check to see if they are breathing. If not, begin rescue breathing. If the heart has stopped, a trained person should give cardiopulmonary resuscitation (CPR).

Name ______________________________

Date ______________________________

TORNADOES
REVIEW 3

Get the Facts

1. Which corner of your basement is the safest to be in during a tornado (unless it has windows)?

2. If your home has no basement, where is the best shelter from a tornado?

3. What should you do if you are in a mobile home during a tornado warning?

4. Where should you go during a tornado if you are in a multistory office building and cannot get to a lower floor?

5. What is the best thing to do if you are in your car and spot a tornado coming?

6. What two ways are used by the weather service to search for tornadoes?

7. Why should you avoid using the telephone or computer during a thunderstorm?

8. What should you do if you are caught in a large open area during a thunderstorm with no shelter nearby?

9. If you are in the woods during a thunderstorm, where would be the best place to seek shelter if there are no buildings nearby?

10. What is the first thing you should do if a person has been struck by lightning?

What Do You Think?

11. Scientists have many unanswered questions about how tornadoes act. Why do you think tornadoes are so difficult to study?

12. Lightning is one force of nature against which we have protection that works. Yet nearly twice as many people are killed by lightning each year as by tornadoes. Why do you think this is true?

Additional Information

Major Tornadoes in the United States

The following is a partial listing of some major tornadoes that have occurred in the United States since the year 1840. A world almanac can give you a more complete list.

Year	Location	Deaths
1840	Natchez, Mississippi	380
1896	St. Louis, Missouri	320
1924	Great Lakes	100
1925	Tri-State Tornado (Missouri, Illinois, Indiana)	689
1932	Alabama	268
1936	Mississippi and Georgia	658
1953	Waco, Texas	114
1965	Six states (Iowa to Ohio)	275
1971	Mississippi Delta	110
1974	148 tornadoes in five states (see pages 45–46)	350
1985	Ohio Valley	90
1990	50 tornadoes in seven states	13
1991	70 tornadoes in one day (seven midwestern states)	30
1994	Tornadoes from Mississippi to North Carolina	45

Activities for Students

1. Imagine that you live in "Tornado Alley." Your family has a special storm cellar in the backyard. List the supplies you want to have on hand for your family. Put an asterisk (*) by the five most important items.
2. Read Frank Baum's *The Wizard of Oz*. Write a summary of the book. How realistic are the accounts of the tornado in this book?
3. Find out about Frank Baum's life. What motivated him to write *The Wizard of Oz*?
4. Visit the local Red Cross office. Ask for free pamphlets about tornadoes and other types of natural disasters.
5. Ask a Red Cross official to speak to the class about natural disasters that could occur in your area.
6. Report on a tornado, severe thunderstorm, or hailstorm that occurred in your area.
7. Look up the origin of the word *tornado*.
8. Read about Benjamin Franklin's famous kite experiment.
9. Find out how thunderstorms affect air travel. Look for information on turbulence, downdrafts, and wind shear as causes of accidents.
10. Report on a famous tornado. Some tornadoes are listed on page 59. For a more complete list, consult a world almanac.
11. Find out more about lightning: its cause, its effects, and attempts to control it.
12. Find out more about waterspouts.
13. Find out how lightning causes forest fires and power outages.
14. Invite a television weather forecaster to speak to the class about severe weather patterns in your area.
15. Write to one of the following for information on severe storms and ways to prepare for them.

    Federal Emergency Management Agency (FEMA)
    National Hurricane Center (NOAA)
    National Severe Storms Forecast Center (Kansas City)
    National Weather Service
    American Red Cross

Tornado Vocabulary

atmosphere	gases surrounding the earth; the air
CPR	cardiopulmonary resuscitation; a way to restore the heartbeat
counterclockwise	in a circle that is opposite to the movement of hands on a clock
cumulonimbus cloud	storm cloud that may generate thunderstorms, tornadoes, or hail
cumulus cloud	white, fluffy cloud
cyclone	tornado
disaster supply kit	kit containing basic necessities like food, water, first aid needs
Doppler radar	radar that can detect tornadoes as they are forming
downdraft	downward-flowing cold air currents associated with thunderstorms
F5 tornado	strongest level of tornado
Fujita-Pearson Tornado Intensity Scale	scale that ranks tornado as to wind speed and damage
funnel cloud	narrow cloud of a tornado that extends down from a cumulonimbus cloud
lightning	natural electricity that flashes between clouds and the ground
manufactured home	mobile home
mesocyclone	updraft that has begun rotating; stage before a tornado forms
National Severe Storms Forecast Center	center where NOAA meteorologists monitor North American weather systems
NOAA	National Oceanic and Atmospheric Administration, a division of the National Weather Service
power surge	strong surge of electricity
rescue breathing	mouth-to-mouth resuscitation
supercell	severe thunderstorm that may generate tornadoes
thunderstorm	storm generating thunder and lightning
tornado	violently swirling windstorm
tornado warning	bulletin warning that tornadoes have been sighted in the area
tornado watch	bulletin warning that conditions are right for tornadoes to form
updraft	column of rising air inside a storm cell

Teacher's Guide for pp. 63–64

This teacher material covers the following reading selection:
Face to Face with Killer Camille

Vocabulary (See page 75 for definitions.)

debris
evacuate
hurricane warning
hurricane watch
velocity

Answer Key (Review 1, p. 64)

1. Dakar, West Africa
2. A hurricane watch and later a hurricane warning were issued. Hurricane flags were flown. The police chief stopped people and warned them to evacuate.
3. a storm surge

Answers for the following will vary.

4. Many people who had experienced other hurricanes thought they could ride out this storm. Other people did not realize the strength of a hurricane.
5. Warnings to evacuate are issued to save lives. But if some people refuse to go, that is up to them.

Additional Information About Hurricane Camille

- Hurricane Camille was the strongest hurricane ever to strike the United States. It is one of only two Category 5 hurricanes on record in this country. The other Category 5 struck the Florida Keys in 1935.
- After Camille finished with the Gulf Coast, it moved inland. When it crossed the mountains of West Virginia, it dumped 2 feet of rain in just a few hours. Massive flooding occurred, causing 155 deaths. Camille finally blew itself out on August 22—17 days after its birth in the Atlantic Ocean.

Face to Face with Killer Camille

It was August 5, 1969. On the Mississippi Gulf Coast, people were enjoying the beautiful weather. It was hot and clear. The beaches were packed with midsummer vacationers.

Meanwhile, across the ocean, a huge band of storm clouds was forming. The clouds formed over the tropical waters of the Atlantic Ocean near Dakar, West Africa. Slowly they began to drift toward the west.

Hurricane Warning Signals

Gale Warning (39–54 mph winds)	Storm Warning (55–73 mph winds)	Hurricane Warning (winds over 73 mph)

By August 14, the storm had reached the Grand Cayman Islands, about 480 miles south of Miami. That morning, a Navy Hurricane Hunter plane flew through the storm, clocking its winds at 60 miles per hour.

By the afternoon of August 15, the winds had reached 115 miles per hour. The storm had become a hurricane. It was the third hurricane of that season. The National Weather Service named it Hurricane Camille.

For the next two days, Hurricane Camille hung around the Gulf of Mexico gaining strength. Then it headed straight for the Mississippi Gulf Coast.

Early on August 16, weather officials issued a hurricane watch all along the coast. By the morning of August 17, they issued hurricane warnings for the entire Mississippi coast. They hoisted red and black hurricane warning flags. They called for immediate evacuation of coastal areas. Thousands of people listened to the warnings. Most of them fled inland. But some did not.

Police Chief Jerry Peralta drove along the highway in Pass Christian, Mississippi. He was looking for people who had ignored the warning to evacuate. He stopped everyone he saw and urged them to leave.

Peralta stopped at the newly built beachfront Richelieu Apartments. There he found Mary Ann Gerlach and 22 of her neighbors. They were having a hurricane party. They had decided not to leave. They were sure they would be safe on the second floor of the luxury apartments.

Storms are a way of life along the Gulf Coast. So the hurricane warnings did not worry the merrymakers. But this was no ordinary storm. It was Killer Camille. It was the strongest hurricane to hit the United States since record keeping began. Later that day, a storm surge crashed into the Richelieu Apartments. The first floor of the building collapsed. Water poured into the second floor. The walls cracked. Mary Ann Gerlach told her husband she was taking her chances outside. She went out the window into the water.

The next thing she remembered was hearing the screams of her husband and friends as they drowned. Then she saw the Richelieu Apartments slip into the water. She could see hurricane lamps lit for the party still glowing. Hours later, the waters dropped her on a pile of debris 5 miles inland from the Richelieu. Of the 23 partygoers, only she had survived.

For more than four hours, Hurricane Camille battered the Gulf Coast with 200 mph winds. More than 132 people died that day. The property damage was almost beyond belief. It was estimated at $1.5 billion.

Name ______________________________

Date ______________________________

HURRICANES
REVIEW 1

Get the Facts

1. Where did Hurricane Camille form?

2. How were people warned that the hurricane was on its way?

3. What force actually destroyed the Richelieu Apartments?

What Do You Think?

4. Residents of the Gulf Coast of Mississippi had plenty of warning to evacuate before Hurricane Camille struck in 1969. Why do you think many people chose to ignore the warnings?

5. Do you think everyone should be forced to evacuate low-lying areas in the event of a hurricane warning? Why or why not?

Teacher's Guide for pp. 66–68

This teacher material covers the following reading selections:

How Are Hurricanes Formed?
How Do Hurricanes Kill?

Vocabulary (See page 75 for definitions.)

baguio	low-pressure area
condense	satellite
cyclone	storm surge
evaporate	typhoon
eye of hurricane	willy willy

Answer Key (Review 2, p. 68)

1. baguio (Philippines); cyclone (Indian Ocean); typhoon (China Seas); willy willy (Australia)
2. off the west coast of Africa; in the Caribbean Sea; in the Gulf of Mexico
3. Water evaporates from the warm ocean. The warm air rises. Cool air rushes into the resulting low-pressure area, whirling upward. As the rising air cools, water vapor condenses, forming clouds around the eye.
4. 74 mph or greater
5. so that large amounts of water evaporate into the air
6. wind; floods; storm surge
7. storm surge
8. The eye is calm. The sun may be shining.
9. They are coming from the opposite direction and often destroy objects weakened by the first winds.
10. It is huge amounts of water that have been sucked up by the hurricane and carried along with it. It may be 30 feet high when it reaches the land.

Answers for the following will vary.

11. People can protect themselves to a degree from high winds by seeking appropriate shelter. There is no protection from a 30-foot-high wall of water due to the enormous force it exerts as it batters into everything in its path.
12. Many people do not tie down their mobile homes correctly. Even a mobile home that is correctly anchored is much more vulnerable to being blown over or crushed than a permanent structure.

Additional Information on Paths of Hurricanes

This map shows paths usually taken by hurricanes. Using a large world map, show students where hurricanes are most likely to hit.

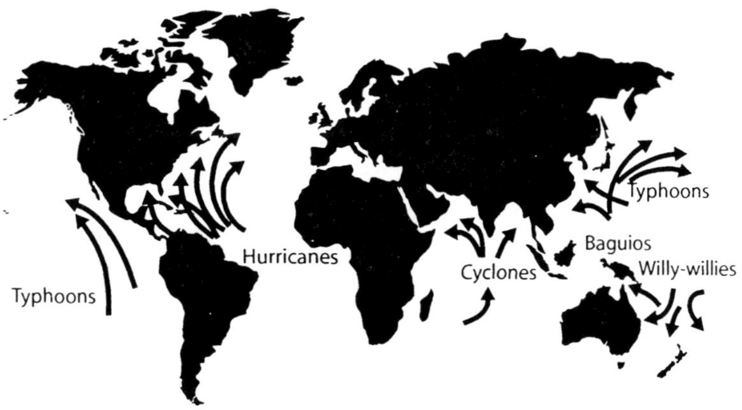

How Are Hurricanes Formed?

The word hurricane comes from the Spanish word *huracán*, meaning evil spirit. The word huracán comes from the name of the Mayan storm god *Hunraken*. Ancient tribes believed that the gods sent hurricanes to show great displeasure with the people. Around the world, there are other names for a hurricane. In the Philippines it is known as a *baguio*. In the Indian Ocean, it is a *cyclone*. In the China Seas it is a *typhoon*. And in Australia, it is a *willy willy*. Whatever the name, the storms are the same.

Hurricanes that hit North America are usually born in the warm waters off the west coast of Africa. They may also start in the Caribbean Sea or the Gulf of Mexico.

In the late summer and fall, the ocean is very warm. The air over the ocean is full of moisture. These are the conditions hurricanes must have to form. The main hurricane season runs from July to October.

The sun shining on the ocean causes much water to evaporate into the air. A current of warm, moist air rises. This makes a low-pressure center on the surface below. Cooler air rushes into this low-pressure area and whirls upward. This makes a whirlpool of air that spirals in and upward in the center. As the warm, moist air rises, it cools. As it cools, the huge amounts of water vapor condense and form gigantic clouds all around the calm central eye.

Sometimes, these clouds break up. Then a thunderstorm results. But other times, the storm continues to grow. It moves over the ocean picking up more water vapor. Its winds increase.

Many tropical storms die at sea. This happens when the storm travels over cooler water or a land mass. But other storms continue to build strength. When the winds reach 74 mph, the storm has become a hurricane. Still it can grow. Its winds can reach up to 200 mph.

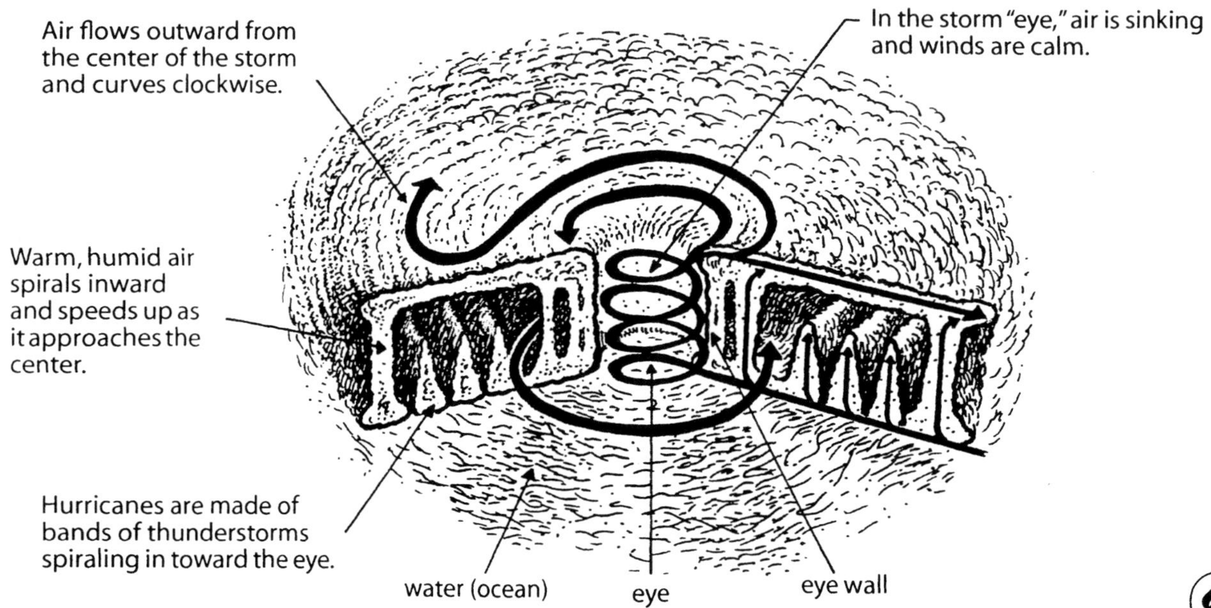

How Do Hurricanes Kill?

Hurricanes carry three forces of destruction: wind, flood, and storm surge.

Wind

Satellite photographs of hurricanes show masses of clouds spinning around a quiet eye in the center. In the eye, the winds are calm. The sun may even shine. But in the walls around the eye, the winds whirl at speeds of up to 200 miles per hour. As the eye passes over an area, the storm appears to be over. But the worst is yet to come. Soon the hurricane strikes again. The winds on the second side come from the opposite direction. Objects damaged by the first winds are often destroyed by the second winds.

At 200 mph, hurricane winds have awesome power. No person can walk in such a wind. Few buildings can withstand these winds. Mobile homes are the first to go. Toppled trees crush people and buildings. Everyday objects such as bicycles or tools become weapons of destruction as they fly through the air at great speed.

Flood

Hurricanes carry huge amounts of water in their swirling clouds. As they move inland, possibly far from the sea, they begin to dump their water. The water in streams rises. Great flooding may occur. Hurricane Camille held its rains until it had traveled all the way from the Gulf Coast of Mississippi to Virginia. There it dumped 2 feet of rain in just 8 hours; 155 people died from the flash flooding that resulted.

Storm Surge

The eye of a hurricane may be 25 to 50 miles across. Inside the eye is a huge low-pressure area. This area serves as a straw. It sucks up vast amounts of water from the ocean. This water is carried along by the hurricane as it reaches land. This is called the storm surge.

As the storm approaches land, billions of tons of water begin to pile up under the eye. When it hits the land, the storm surge may be 30 feet high. It engulfs everything in its path. It may continue its roll of destruction 20 or more miles inland. Most deaths in hurricanes are caused by storm surges. The only protection is to get away from it.

Wind, flood, and storm surge. Combined, these make hurricanes a deadly force of nature.

Name ______________________________

Date ______________________________

HURRICANES
REVIEW 2

Get the Facts

1. What are some other names for hurricanes?

2. Where do most hurricanes that hit North America start?

3. Describe how hurricanes form.

4. How fast must winds be for a storm to be called a hurricane?

5. Why must the ocean water be warm for a hurricane to form?

6. Name the three forces of destruction brought by a hurricane.

7. Which of these three forces causes the most deaths in a hurricane?

8. Why does the eye of the hurricane fool many people into thinking the storm is over?

9. Why are the winds on the second side of the hurricane so destructive?

10. What is a storm surge?

What Do You Think?

11. Why do you think most hurricane deaths are caused by drowning rather than by high winds?

12. Why are mobile homes so dangerous in storms?

Teacher's Guide for pp. 70–72

This teacher material covers the following reading selections:

How Can Hurricanes Be Controlled?

How Can You Get Ready for a Hurricane?

Vocabulary (See page 75 for definitions.)

hurricane categories
radar
meteorologists

Answer Key (Review 3, p. 72)

1. They use information gathered by radar, weather satellites, and weather planes.
2. A hurricane is possible in the area within the next 36 hours.
3. A hurricane is expected to hit the area within the next 24 hours.
4. Listen to radio or TV for further instructions. Do what you can to protect your property from damage by high winds. Evacuate immediately if told to do so. If you do not evacuate, go to a center room away from windows (closet or windowless bathroom).
5. by their wind velocity, in increasing order from Category 1 to Category 5
6. Category 5
7. There is an alphabetical list for each year.

Answers for the following will vary.

8. first aid kit; medications needed on a regular basis; battery-powered flashlight; battery-powered radio or TV; extra batteries; sleeping bags or other bedding; rain gear; extra clothes; canned food; can opener; containers for drinking water; other special items needed by infants, elders, or disabled family members
9. family heirlooms; photographs; cash; valuable jewelry, etc.

Additional Information on Hurricane Names

Since 1953, hurricanes have been given official names by the National Weather Service. At first, they were given only women's names. But since 1979, men's names have also been included on the list.

There are six different lists of names for hurricanes. One list is used for each year. At the end of six years, the first list is used again. On each list, the names are in alphabetical order. The first storm of the year will be given the name beginning with the letter A. The next storm will be given the name beginning with B, and so on. The letters Q, U, X, Y, and Z are not used.

There are separate lists of names for hurricanes in the Pacific Ocean. In Asia, hurricanes are usually given a number rather than a name.

How Can Hurricanes Be Controlled?

Nobody can stop a hurricane. In the past, people were at the mercy of hurricanes that seemed to strike with little warning. But today, most deaths from hurricanes can be prevented, thanks to early warnings by the National Weather Service.

Since the 1960's, weather satellites have been used to track storms. Radar-equipped planes guided by satellite photographs can fly directly to the storms to get more information. Flying directly into the hurricane, scientists can measure wind speed and direction as well as the amount of water carried by the storm. Meteorologists then use the information gathered by satellites and planes to predict when and where a hurricane will strike.

The National Weather Service issues bulletins to advise people about a storm's strength and location. It advises people what actions to take to safeguard themselves and their property. In the case of Hurricane Camille, it was estimated that the National Weather Service saved at least 50,000 lives by warning people of the impending danger.

Meteorologists classify hurricanes on a scale of 1 to 5 by their wind speeds. The greater the wind speed, the higher the number and the greater the destruction of the storm.

Category 1: 74 to 95 mph

Category 2: 96 to 110 mph

Category 3: 111 to 130 mph

Category 4: 131 to 155 mph

Category 5: more than 155 mph

In the last century, only two Category 5 storms have hit the United States. In 1935, one hit the Florida Keys. And in 1969, it was Hurricane Camille.

Meteorologists are also responsible for naming each hurricane. They use an alphabetical list of male and female names. The first hurricane of the year gets a name beginning with A; the next gets a name starting with B, and so on. No hurricanes are named with names starting with Q, U, X, Y, or Z. That leaves a list of 21 names for each year. So far, scientists have never needed more names than that.

How Can You Get Ready for a Hurricane?

Before a Hurricane Comes

The United States is hit by hurricanes about twice a year. Much of the coast is hurricane country. Find out if you live in a hurricane-prone area. Find out what to do to prepare for hurricanes.

Decide ahead of time where you would go in case of a hurricane. This could be the home of a friend who lives inland. It could be a hurricane shelter. Or it could be a motel inland. Be sure you know how to get to your emergency shelter.

As hurricane season approaches, be sure your disaster supply kit is in order. Include these items:

- first aid kit
- any medications needed on a regular basis
- battery-powered flashlight and radio; extra batteries
- sleeping bags or bedding
- rain gear
- canned food; can opener
- containers for drinking water

Know what the terms *hurricane watch* and *hurricane warning* mean. A hurricane watch means that a hurricane is *possible* in the area within the next 36 hours. A hurricane warning means that a hurricane is *expected* to occur in the area within the next 24 hours or less.

If a hurricane watch is given, stay tuned to your radio or TV to stay up-to-date on the situation. Keep your car filled with gas, in case you need to leave the area. Make sure you have batteries, canned food, drinking water, first aid kit, and any needed medications on hand.

If a hurricane warning is given, listen to your radio or TV for instructions. If you are told to evacuate, do so. If you are not advised to evacuate, stay inside away from windows when the storm approaches. Choose a spot in the center of your home in a closet or bathroom without windows.

You can prevent some damage from high winds. Cover the windows of your home with shutters or plywood. Bring inside any items that can be picked up by the wind: garbage cans, lawn furniture, hanging plants, etc. If you live in a mobile home, check its tie-downs. Try to find a permanent structure in which to ride out the storm.

When the storm is over, listen to the radio or TV to learn when it is safe to return home.

Name ______________________________

Date ______________________________

HURRICANES
REVIEW 3

Get the Facts

1. How do meteorologists predict what a hurricane will do?

2. What is a hurricane *watch*?

3. What is a hurricane *warning*?

4. What should you do in the event of a hurricane warning?

5. How are hurricanes classified?

6. In what category was Hurricane Camille classified?

7. How are hurricanes named?

What Do You Think?

8. Imagine that you live in a hurricane-prone area. As hurricane season approaches, you prepare a disaster supply kit. You will use this kit if a hurricane leaves you stranded in your home for several days. List 10 things that you would be sure to include in the kit. List them in order of importance (most important first).

9. A hurricane warning has been issued for the area you live in. You have been ordered to evacuate. You can only take a few things with you. Other than your disaster supply kit, what personal items would you take with you?

Additional Information

Ten Hurricanes Causing the Most Deaths in the World		
Year	**Location**	**Deaths**
1970	Bangladesh	500,000
1881	Indochina	300,000
1876	India	200,000
1882	India	100,000
1864	India	70,000
1942	India	40,000
1965	Bangladesh	25,000
1963	Bangladesh	22,000
1960	Bangladesh	14,000
1965	Bangladesh	13,000
1906	Hong Kong	10,000

India has often been struck by hurricanes roaring up through the Bay of Bengal. The Ganges-Brahmaputra delta at the head of the Bay of Bengal is one of the most often struck areas in the world. Another is Bangladesh (formerly East Pakistan). Both of these places are very flat, low-lying areas. They are also some of the most densely populated areas on earth—more than 1400 people per square mile. They are very poor areas. And they lack modern warning systems. Thus, when storm surges of 10 to 30 feet sweep over the flatland, there is no place to hide and no time to evacuate. This accounts for the extremely high death tolls.

United States Hurricanes

Florida is the state most prone to hurricanes. It is struck by one or more hurricanes almost every year. However, other coastal states are vulnerable as well.

The deadliest hurricane to strike the United States hit Galveston, Texas, in 1900. More than 6000 of the total 38,000 residents were killed. A huge storm surge completely flooded the city. To protect Galveston from future storm surges, a 17-foot-high hurricane wall was constructed along the beachfront.

The following are some of the deadliest U.S. hurricanes.

Year	Location	No. of Deaths
1900	Galveston, Texas	6000
1928	Okeechobee, Florida	1836
1979	Eastern U.S. & Caribbean (David)	1100
1919	Florida Keys and Texas	800
1938	New England	600
1935	Florida Keys (Labor Day Hurricane)	408
1957	Louisiana (Audrey)	390
1909	Mississippi and Louisiana	350
1969	Mississippi, Virginia (Camille)	256
1980	Texas and Caribbean (Allen)	272
1989	Southeast United States and Caribbean (Hugo)	504
1955	New England (Diane)	400
1972	Florida to New York (Agnes)	118
1965	Florida & Louisiana (Betsy)	75
1992	Florida (Andrew)	30

Activities for Students

1. Make up your own list of names for hurricanes. Do not use the letters Q, U, X, Y, or Z.
2. Devise a new system for identifying hurricanes.
3. Find out about storm flags used by ships. What flags are used to warn of an approaching hurricane? Draw a set of these warning flags.
4. Contact the local Red Cross or Federal Emergency Management Office for more information about hurricanes. These agencies have many pamphlets available. Ask a representative from one of these organizations to speak to your class about emergency preparedness appropriate for your area.
5. Research one of the deadly U.S. hurricanes. Share your information with the class.
6. Write an eyewitness account of what you think it would be like to ride out a hurricane in your home or in a shelter.

Hurricane Vocabulary

baguio	Philippine word for hurricane
condense	change from water vapor back to liquid form
cyclone	name for hurricane in the Indian Ocean
equator	imaginary line around the middle of the earth halfway between the north and south poles
evacuate	to leave an area considered dangerous
evaporate	when a liquid (or solid) turns into a gas
eye	the center of a hurricane
hurricane	strong storm with winds over 73 mph
hurricane warning	hurricane is expected in the warning area in the next 24 hours
hurricane watch	hurricane is possible in the watch area within the next 36 hours
low-pressure area	area of low barometric pressure, usually associated with cloudy, rainy weather
meteorologist	scientist who studies weather
radar	instrument that uses radio waves for telling the direction and speed of unseen objects
satellite	manufactured object that orbits earth, sending back information
storm surge	sudden onrush of very high water during hurricane
tropical	in the warm areas near the equator
typhoon	name for hurricanes in the China Seas
velocity	speed
water vapor	water that has evaporated into the air and is suspended in the air
willy willy	Australian name for a hurricane

Teacher's Guide for pp. 77–78

This teacher material covers the following reading selection:
The Buffalo Blizzard

Vocabulary (See page 89 for definitions.)

blizzard	windchill factor
blizzard warning	zero visibility
visibility	

Answer Key (Review 1, p. 78)

1. Arctic winds from Canada cross Lake Erie. Over the lake, they pick up a lot of moisture. That moisture falls on land as snow. The city of Buffalo is on the east end of Lake Erie. That is where most of this snow is unloaded.
2. very cold with lots of snow
3. winds of 68 mph with gusts up to 80 mph; blowing snow; 10° below zero with a windchill factor of 60° below
4. Telephone lines were down. No one could get out or in by car, plane, or train.
5. The city could get government help with the cleanup.

Answers for the following will vary.

6. Avoid going out in storms. Have emergency supplies such as food, batteries, and first aid items on hand. Have adequate clothing for cold weather.

Additional Information on the Buffalo Blizzard

- January 1977 was the coldest single month on record in the eastern United States. Most areas averaged 12° to 18° below normal.
- The blizzard that struck Buffalo also struck many central and eastern states. Schools, factories, and businesses closed from Wisconsin to Georgia.
- Lake-effect snowstorms occur to the lee of the Great Lakes. Arctic air picks up moisture as it swoops over the lakes. Buffalo and Watertown, New York, are periodically covered with snow from lake-effect storms.
- Buffalo ended that winter with a record 199 inches of snow. Boonville had 346 inches; Barnes Corners had 370 inches; Old Forge had 408 inches and Hooker had 466 inches (39 feet). These towns lie to the lee of Lake Ontario.
- The winter of 1978 was even more extreme in much of the eastern half of the United States. The winters of 1977 and 1978 together were the two coldest consecutive winters on record in the United States.
- These cold winters intensified the migration of many northerners to the warmer states of the Sun Belt.

The Buffalo Blizzard

People living in Buffalo, New York, are used to a lot of snow. Buffalo is located on the east end of Lake Erie. When arctic winds sweep down from Canada, they pass over the Great Lakes. They pick up huge amounts of moisture from the lakes. When they hit land, they dump this moisture as snow. Buffalo often gets 3 feet of snow in a single storm. But the storm that struck on January 28, 1977, had never been equaled in Buffalo.

Buffalo was having a hard winter. Already the city had received 12.5 feet of snow for the season. And the temperature had not made it above freezing since Christmas.

Things were about to get worse. At 5:00 A.M. the Buffalo Weather Service issued a near-blizzard storm warning for that afternoon. By 11:00 A.M., the weather service changed it to a blizzard warning. The blizzard struck as the warning was being announced.

Winds of 68 mph with gusts up to 80 mph blew tremendous amounts of snow into town. The temperature dropped to 10 degrees below zero. The windchill factor hit 60 below.

Visibility dropped to zero. It would remain at zero for the next 25 hours. Snow began piling up in drifts. Some drifts reached 25 feet high.

Most downtown workers were stuck in the city. Many who tried to drive home got trapped in their cars. Others found shelter in public buildings.

The storm cut off the city from the outside world. It downed telephone lines. It halted cars, trains, and planes. It even buried the snowplows.

Businesses, schools, and government offices closed. Shortages of food, gasoline, and heating oil developed. People came to each other's aid. They shared what they had.

President Jimmy Carter declared a state of emergency on January 29. This allowed the government to give aid to the city.

But it was fully a week after the blizzard before massive amounts of aid reached Buffalo. First, soldiers and snowplows were flown in from military bases to help clear the streets.

The National Guard was called in to use bulldozers and tanks to cut through the drifts. Army helicopters arrived to assist in rescue operations. Police made their rounds on snowmobiles.

There was really no place to put all the snow. So some of it was loaded onto open freight cars. As these snow trains moved south, the snow gradually melted and disappeared.

The Buffalo blizzard claimed 29 lives. It cost $300 million in lost sales and wages. And it heavily damaged many homes and cars. In all, Buffalo had 199 inches of snow that year. (The previous record snowfall of 124 inches had stood for 67 years.)

Name ______________________________

Date ______________________________

Winter Storms
Review 1

Get the Facts

1. Why does Buffalo, New York, get a lot of snow each winter?

2. What had the winter been like before the blizzard struck?

3. Describe the conditions during the blizzard.

4. Why was Buffalo cut off from the outside world after the blizzard?

5. What did it mean for the city when President Carter declared a state of emergency?

What Do You Think?

6. What precautions could people in northern cities take to avoid injury or death in severe winter storms?

Teacher's Guide for pp. 80–82

This teacher material covers the following reading selections:

What Is a Blizzard?

Other Winter Hazards

Vocabulary (See page 89 for definitions.)

avalanche	sleet
blizzard	visibility
blizzard warning	wet avalanche
drift	windchill factor
ice storm	zero visibility
powder avalanche	

Answer Key (Review 2, p. 82)

1. a winter storm involving high winds, blowing snow and low temperatures
2. You wouldn't be able to see anything but white.
3. It tells the combined effect of temperature and wind on the bare skin.
4. Raindrops freeze or rain freezes on snowflakes, forming pellets of ice called sleet.
5. when rain has been falling and the temperature drops below 32°F
6. They turn streets into sheets of ice. Trees and power lines fall under the weight of the ice.
7. Wet-snow avalanches are more powerful because wet snow is much heavier than powder snow.
8. Wet-snow avalanches are more predictable. Therefore, people can stay out of the areas where they are likely to occur. People can avoid building there. Also, protective barriers can be built.
9. (any three) Conditions near populated areas can be monitored. Warnings can be issued if conditions are right for avalanches. Explosives can be detonated to clear snow before it builds to dangerous levels. Barriers can be built in areas that are avalanche-prone. Protective tunnels can be built over roads or railroad tracks in dangerous areas.
10. Many victims are killed as the avalanche occurs. Those left alive must usually be rescued within an hour before they run out of air to breathe.

Answers for the following will vary.

11. Today, many more people are using the mountains for recreation. They go cross-country skiing, snowshoeing, hiking, climbing, and camping. Also, more people are building in mountain areas.
12. Check with authorities in the area. Find out where the danger of an avalanche exists. Stay out of that area and others that have been posted. Make sure people know where you will be skiing.

Additional Information on Blizzards and Cold Weather

1. There are about 100 snow-related deaths in the United States each year. Most of these deaths occur in blizzards.
2. Most blizzards occur in the middle latitudes of North America, Europe, and Asia.
3. It is difficult to tell when a snowstorm becomes a blizzard. The National Weather Service issues a blizzard warning when winds reach 35 and blow the snow about, causing low visibility. The weather service issues a severe blizzard warning when winds reach 45 mph, with a lot of snow, and temperatures below 10°F.
4. The combination of strong winds and low temperatures causes hypothermia (low body temperature) very quickly. A person in wet clothes will suffer hypothermia much faster than one who is in dry clothes.

What Is a Blizzard?

A blizzard is the most dangerous kind of winter storm. A blizzard has strong winds, blowing snow, and low temperatures. The National Weather Service issues blizzard warnings when winds of 35 mph or more blow large amounts of snow, creating poor visibility.

Dangers in a blizzard include low visibility, snowdrifts, low temperatures, and windchill.

Low Visibility

As the strong winds of a blizzard blow snow about, visibility becomes very limited. In some cases, visibility drops to zero. When this happens, people can easily lose their way. They can even get lost trying to walk a very short distance. There have been many cases of farmers getting lost on the way to the barn and freezing to death only steps away from safety.

Snowdrifts

As large amounts of snow are blown about, huge drifts pile up. These drifts can quickly bring all traffic to a halt. Vehicles can get stuck in the drifts. And people can get trapped in the vehicles—sometimes for days.

Snow can also be very heavy. Ten inches of snow on the roof of a 1500-square-foot house weighs about 40 tons. Such weight can cause roofs to cave in and buildings to collapse. Also, it is also a lot of work to shovel large amounts of snow!

Low Temperatures

In blizzard conditions, the temperature is low. There is no specific temperature. But it is very cold.

Windchill

The wind makes a difference in how cold we feel in winter weather. For example, a temperature of 30°F with 45 mph winds feels colder than a temperature of 15° with no wind.

Meteorologists have developed a windchill table. This table tells how cold the temperature and wind combined will feel to bare skin.

Other Winter Hazards

Ice Storms and Sleet

When raindrops freeze or when rain freezes on falling snowflakes, pellets of ice form. The ice pellets are called sleet. Sleet makes roads and sidewalks very slippery. These slippery conditions pose a danger to cars and pedestrians.

An ice storm can shut down an entire city for days. This type of storm takes place when temperatures drop below 32°F after rain has been falling. The water freezes. The streets turn into sheets of ice. Trees and power lines often fall under sheer weight of the ice.

Avalanches

Avalanches are massive slides of snow, ice, rocks, or earth. Avalanches hurl down a mountainside at speeds of up to 200 mph. They can be triggered by a loud noise like a gunshot. They can be started by a person or animal crossing a slope at the wrong place. They also can be set off by a heavy snowfall or melting snow.

There are two types of snow avalanches: powder and wet. A powder avalanche is made up of loose, dry snow. A wet-snow avalanche is much more powerful. It often carries boulders and trees down with it. Usually it forms in the spring and follows a known path. Because they are more predictable, wet avalanches cause less damage than the unpredictable powder avalanches.

Chances of rescue or survival for anyone caught in an avalanche are not good. Many victims die as the avalanche occurs. Most of those who are still alive have to be rescued within an hour before they run out of air to breathe. Only a very few victims are rescued in time.

About 100,000 avalanches take place every year in the Rocky Mountains. Most are in isolated areas and do no harm to people. But deaths from avalanches are on the rise as more people take up winter sports such as skiing and snowshoeing.

Much avalanche danger can be avoided. Conditions near populated areas or resorts can be monitored. And avalanche warnings can be issued if dangerous conditions exist. Explosives can be used to clear snow before dangerous buildups occur. Barriers can be set up to divert avalanches from buildings. Protective tunnels can be built over sections of road or railroad track in dangerous areas.

Most of the time, avalanches can be predicted. Known avalanche paths can be monitored. People can be educated about areas to avoid when skiing or climbing.

Name ______________________________

Date ______________________________

Winter Storms
Review 2

Get the Facts

1. What is a blizzard?

2. When there is zero visibility in a blizzard, what does it look like outside?

3. What is the windchill factor?

4. What causes sleet to form?

5. When do ice storms occur?

6. Why are ice storms so dangerous?

7. Which type of avalanche is more powerful? Why?

8. Which type of avalanche is more predictable and, therefore, causes less damage to buildings and people?

9. Name three ways avalanche danger can be lessened.

10. Why are so few people buried by an avalanche rescued alive?

What Do You Think?

11. Our means of predicting avalanches has greatly improved. Yet deaths from avalanches in the Rocky Mountains and the Swiss Alps are on the increase. Why do you think this is true?

12. You are planning a cross-country skiing trip in the Rocky Mountains. What should you do before you go to avoid avalanche danger?

Teacher's Guide for pp. 84–86

This teacher material covers the following reading selections:

How Does Severe Winter Weather Affect People?
How Can You Keep Safe in Winter Weather?

Vocabulary (See page 89 for definitions.)

carbon monoxide	mild hypothermia
exhaustion	overexertion
frostbite	severe hypothermia
hypothermia	winterized

Answer Key (Review 3, p. 86)

1. subnormal body temperature caused by prolonged exposure to cold
2. Wrap the victims in warm blankets or clothing. Get them into a warm place. Call the doctor if heart rate seems slow.
3. the very old and very young
4. numbness; white or grayish-yellow skin patches
5. Warm the area gently with your hands or with warm—not hot—water. Take the person to a doctor.
6. (any three) Have emergency supplies on hand. Winterize your car. Have your furnace checked. Have warm clothing on hand.
7. Layers of clothing will keep you warmer. Wear a hat and gloves.
8. You tire more easily. About a third of deaths from winter storms occur from overexertion. Victims are usually older people.
9. Tie something colored to the antenna. Avoid running the engine with the windows closed. Exercise a little to keep warm. Check to be sure the exhaust pipe is not blocked by snow.
10. You can easily get lost in a blizzard. Rescuers can find you more easily in a car. The car provides some shelter and warmth.

Answers for the following will vary.

11. Snow is very heavy. Shoveling large amounts of snow is a big strain. It puts a great deal of stress on the heart of someone who is not used to exercise. Many older people suffer heart attacks while shoveling snow.
12. One of the symptoms of hypothermia is confusion. Victims of hypothermia become disoriented. They do not realize what is wrong. And they do not know what to do for themselves.

Additional Information on Winter Weather

- Some scientists think the earth's climate is cooling. Some think we may even have another ice age. Cooler winter weather means cooler oceans. Cooler oceans mean more droughts since less moisture evaporates from lower-temperature waters.
- The National Weather Service issues the following kinds of winter weather watches and warnings.

Winter storm watch/warning	a storm may be/is nearing
Heavy snow warning	4 inches of snow may fall in the next 12 hours; or 6 inches in the next 24 hours
Blizzard warning	heavy snow blown by 35 mph winds is expected
Severe blizzard warning	heavy snow blown by 45 mph winds; temperature under 10°F
Ice storm warning	freezing rain expected

How Does Severe Winter Weather Affect People?

Hypothermia

When most people think of deaths from the cold, they think of "freezing to death" or dying "from exposure." Actually, these deaths have been caused by hypothermia.

Hypothermia is low body temperature caused by prolonged exposure to cold. In a mild state of hypothermia, the body temperature drops to 90°–95°F. When this occurs, both the heart and breathing rates slow down. Persons with mild hypothermia are alert but shivering severely. They may feel tired and irritable.

Wrap the victims in warm blankets or clothes. Put them in a warm shelter. If heart rate is slow, call a doctor.

Severe hypothermia occurs when the body temperature has dropped below 90°F. Persons suffering from severe hypothermia cannot think clearly enough to take care of themselves. They should be bundled up warmly and taken to a hospital immediately. If the body temperature drops below 60°F, the heart will stop beating and death will occur.

Those in greatest danger of suffering hypothermia are the very young and the very old. These people are less aware of feeling cold. Their bodies may lose heat quickly and regenerate it slowly.

Frostbite

Signs of frostbite include numbness and skin that looks white or grayish-yellow. Frostbite usually affects feet, hands, and ears.

If frostbite occurs, do not rub the area. Rubbing can damage the skin. Do not expose the area to intense heat as from a heating pad or stove. Instead, warm the area slowly with your warm hand or in warm water (100°F). If blisters occur, do not touch them. Take the person to the doctor or emergency room as soon as possible.

Exhaustion and Overexertion

About a third of the deaths from winter storms occur from exhaustion or overexertion. Many people suffer heart attacks or strokes after shoveling snow or using a snowblower.

Car Accidents

Accidents are the prime cause of deaths during winter storms. Low visibility during a blizzard can cause traffic to pile up. Vehicles can go off the road. And they can become stuck in drifts, trapping passengers inside.

How Can You Keep Safe in Winter Weather?

Before Winter Weather Begins

If you live in a cold-weather area, get ready ahead of time. Have appropriate clothing on hand. Make sure your vehicle is ready for winter driving. (This includes adding antifreeze to the radiator.) Have the furnace in your home checked to be sure it is working properly. Keep an emergency supply of canned foods and extra batteries on hand in case you can't get out for a few days.

During Winter Weather

The National Weather Service warns people when snowstorms, ice storms, blizzards, or severe blizzards are approaching. It issues watches or warnings as needed. A *watch* means that a storm is approaching. A *warning* means severe conditions are expected in the immediate future.

If a winter storm watch is issued, make sure you are prepared. Stay tuned to radio or TV for updated weather conditions.

If a winter storm warning or blizzard warning is announced, take steps at once to protect your life and property.

In severe winter weather, stay home if possible. If you must go out, dress appropriately. Layers of clothing will keep you warmer than one coat. Cover your head and hands to prevent loss of body heat. Cover your mouth to protect your lungs.

If you have to shovel snow or work outside, pace yourself. Work slowly. Then rest to avoid overexertion. This is especially important for older people.

Travel is especially hazardous in severe weather. Drive only if you absolutely must. Make sure someone knows your destination and route. Keep the gas tank as full as possible. Carry emergency supplies in your vehicle. These supplies could include blankets, dry clothes, a shovel, sand, a flashlight, a first aid kit, jumper cables, and tire chains.

If you get caught driving in a blizzard, stop and try to find shelter. If no shelter is available, stay in your car. Do not risk getting lost in a blizzard. Outside, you have no protection from the cold. In your car, you have some protection. And rescuers can find you more easily.

But you do get out of the vehicle to do two things. First tie something bright colored on your antenna. Second, make sure the exhaust pipe is clear of snow, so fumes do not back up into the car. Then, for about 10 minutes every hour, start the car and use the heater. Keep one window on the side away from the wind open a crack to let fresh air in. Don't run the motor with the windows closed. You could die of carbon monoxide poisoning. Keep moving your arms and legs to keep warm.

Name ______________________________

Date ______________________________

WINTER STORMS
REVIEW 3

Get the Facts

1. What is hypothermia?

2. What is the correct first aid for people suffering from mild hypothermia?

3. Which two groups of people are in greatest danger of hypothermia?

4. Name two signs of frostbite.

5. What is the correct first aid for frostbite?

6. Name three things you should do before winter arrives if you live in a cold-weather area.

7. Describe how to dress appropriately if you must go outside in very cold weather.

8. Why is it important to avoid overexertion when working outside in the cold?

9. What should you do if you are trapped in your car during a blizzard?

10. Why is it safer to stay in your car rather than to walk for help if you are stuck during a blizzard?

What Do You Think?

11. Why do you think that one third of deaths from winter storms are caused by overexertion?

12. Why do you think victims of hypothermia may not be able to do the right first aid for themselves?

Winter Storms
Additional Information

Some Winter Weather Disasters

- The Donner party was a group of 100 people traveling from Springfield, Illinois, to California in 1846. They got trapped in the Sierra Nevada Mountains by a blizzard. They were stopped by 3-foot-deep snow and bitter cold. In December, 16 of them tried to walk out. Eight of them died. Of the group remaining in camp, only 48 survived. Those that survived did so by eating the flesh of others who had died.
- In March 1988, a powerful blizzard struck the northeast United States. It killed 400 people, including 200 in New York City.
- The Buffalo, New York, blizzard of 1977 killed nearly 200 people.
- During the winter of 1981–82, a series of blizzards blew their way across the midwestern United States. Snowfall was heavy in much of the country. In California, heavy rain caused massive mud slides. The mud slides killed 30 people. In January, a severe cold wave broke records all across the country. Hundreds of people died from what was called the Arctic outbreak of the century.
- The year 1816 is known as the Year Without a Summer. In April 1815, the volcano Tambora on the island of Sumbawa in Indonesia erupted. This eruption was probably the greatest in total amount of matter erupted in modern times. Some 35 cubic miles of debris formed a thick layer of dust in the atmosphere. Around the world, that dust blocked much sunlight for several years. Cold weather, including frost, snow, and ice continued over much of the United States for the entire summer.
- An ancient winter disaster occurred in the Alps in 218 B.C. Hannibal was crossing the mountains with his army of 38,000 men, 8000 horses, and 37 elephants. Before the crossing was complete, half of the men had been killed in avalanches.
- During World War I (1915–18), avalanches killed 60,000 people in the Italian-Austrian Alps. Some of these avalanches were set off on purpose.
- During the winter of 1942–43, some 260,000 German soldiers died near Stalingrad, Russia, as a result of severe winter weather.
- An avalanche near Huascarán, a mountain in western Peru, killed over 4000 people.

Some Winter Records

Record	Location	Date
Coldest temperature in United States: –79.8°F	Prospect Creek, Alaska	1-23-71
Coldest temperature on earth: –126.9°F	Vostok, Antarctica	8-24-60
Most snow in one year: 1224.5 inches	Mount Rainier, Washington	1971–72
Tallest snowman (Super Frosty): 63.56 feet	Alaska	1988

Activities for Students

1. Ask the National Weather Service to send a meteorologist to talk to the class. Prepare some questions about winter storms and what to do in case you are trapped in one.
2. Read safety tips for driving in icy and snowy conditions.
3. Find out more about one of the following winter weather disasters. Report on your findings to the class.
   - The Donner party
   - The 1888 blizzard in New York
   - The 1981–82 blizzard in the Midwest
   - The Year Without a Summer
   - Hannibal crossing the Alps
   - Avalanches in the Alps
   - Avalanches in World War I
   - The 1942–43 Battle of Stalingrad
   - The avalanches on Huascarán in Peru in 1962 and 1970
   - Valley Forge in 1777–78
4. Find out and report to the class about first aid for hypothermia, frostbite, or heart attack.
5. Read more about the theory that we are entering a little ice age.
6. Read the story of Anna Conrad. Conrad was buried for five days in an avalanche in Squaw Valley, California, in March 1982. She lived to tell about the experience.
7. Read the story of Lenny Skutnik. Skutnik dove into the icy Potomac River to save a victim of an airline crash that occurred during blizzard conditions. Unfortunately, he was unable to save the lives of 78 other people who died in the crash of the Air Florida jetliner.
8. Read about the snowflake studies of Johannes Kepler (1571–1630). Kepler was the first to discover that all snowflakes are six-sided.
9. Find out about the 5200-year-old man who was discovered buried in a glacier in the Alps in 1991. The man may have died in an avalanche. He has been called *Homo tyrolensis* or the Iceman. He is the most ancient human to have been found intact.
10. Switzerland has the most avalanche fatalities in the world. In 1936 the country set up the Swiss Federal Institute for Snow and Avalanche Research. About 70 observatories keep an eye on problem areas. Avalanche warnings are issued as needed. Find out more about avalanche prediction.
11. Steps can be taken to lessen the damage done by avalanches. Design structures that could block or divert the flow of an avalanche coming down a mountainside. Examples of structures that are now used include: planting trees, putting up fences or walls, wedge-shaped structures that divert the snow out to the sides, tunnels over train tracks or roads, buildings shaped like a wedge so snow is diverted around the sides, etc.
12. Read more about glaciers.

Winter Storms Vocabulary

avalanche	massive slides of snow, ice, or rock that hurl down a mountainside at great speed
blizzard	winter storm characterized by winds over 35 mph, low temperature, and blowing snow
blizzard warning	heavy snow blown by winds of 35 mph or more will be arriving soon
carbon monoxide	poisonous, odorless, colorless gas that is emitted in the exhaust of cars
drift	pile of snow that has been heaped up by the wind
exhaustion	total loss of strength
frostbite	injury to the tissues of the body due to freezing
hypothermia	subnormal body temperature caused by prolonged exposure to cold
ice storm	rain that freezes when the temperature drops below 32°F
mild hypothermia	body temperature of 90 to 95°F
overexertion	putting too much physical effort into a task, so as to cause exhaustion
powder avalanche	avalanche of loose, powdery snow
severe hypothermia	body temperature under 90°F
sleet	when rain freezes onto falling snow, forming ice pellets
wet avalanche	avalanche of heavy, wet snow that carries trees, rocks, etc., along with it
windchill	how the temperature feels on exposed skin, figured by combining the wind and temperature
winterized	with a car, prepared for winter weather by having antifreeze added, etc.
visibility	range of possible vision
zero visibility	very little vision is possible

When the Waters Rise

Teacher's Guide for pp. 93–94

This teacher material covers the following reading selection:
The Big Thompson Flash Flood

Vocabulary (See page 111 for definitions.)

Answer Key (Review 1, p. 94)

1. It was a three-day weekend to celebrate the hundredth anniversary of Colorado's statehood.
2. The canyon was very narrow with walls of granite. There was no place for the water to go except straight down the canyon.
3. The flood swept away cars, houses, people, and buildings. It destroyed a power plant. It tore up a highway.

Answers for the following will vary.

4. The people were probably not aware that they were in a flash flood-prone area. If so, they might have evacuated when the river first began to rise.
5. People who lived in the canyon should have known of its past history. But tourists would not realize the danger, unless signs were posted.

Additional Information on Flash Floods

- Flash floods are the number one weather-related killer in the United States today. They cause an average of 200 deaths a year.
- Some 15,000 areas have been identified by the Flood Insurance Administration as prone to flash floods. Many of these areas are in narrow mountain canyons or dry washes. Most people who use these areas for recreation are unaware of the danger.
- A flash flood that occurred in Rapid City, South Dakota, was similar to the Big Thompson flood. In this 1972 flood, 237 people died. The trouble began with a group of stationary thunderstorms. Positioned over the Black Hills, the storms dumped 15 inches of rain in six hours. The rain rushed into two dozen streams. The overloaded streams roared down the 22 miles to Rapid City in just 3.5 hours. A man who saw a wall of water rushing down the creek issued the first alert. Next, radio stations advised residents to evacuate. Then the National Guard warned people. While some people heeded the warnings, others just didn't believe what they were hearing. Many of these people paid with their lives.
- Most flash floods occur in April through September. In the eastern and central United States, they occur mostly at night. In the west, they occur during the afternoon.
- The storms at Big Thompson developed when strong, low-level easterly winds pushed moist, unstable air into the mountains. There were no high-altitude winds to blow the storm clouds away. So the clouds remained and dumped all their rain on the area.
- Flash floods can also occur in desert regions. These areas have soil that is so dry and hardened that water cannot readily soak in. So when a thunderstorm occurs, the water simply rushes along dry ravines and desert washes. Campers should avoid making camp in such areas, particularly if rain is forecast.

Floods and Tsunamis

The Big Thompson Flash Flood

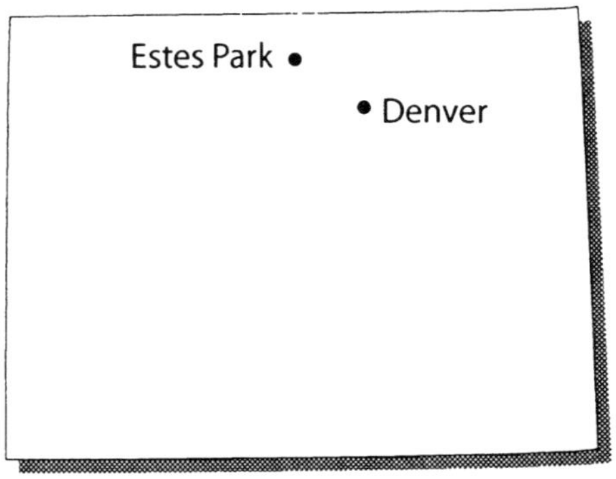

Colorado

July 31, 1976, was a state holiday in Colorado. It was the eve of Colorado Day, the hundredth anniversary of Colorado's statehood. Around the state, thousands of residents and tourists were on a three-day weekend.

Big Thompson Canyon is a scenic area not far from Estes Park, Colorado. Campgrounds and motels in this granite canyon were full of vacationers. People were fishing, hiking, and wading in Big Thompson River.

Big Thompson River is not big at all. It is tiny. In spots it is barely 12 feet across. It lies at the bottom of Big Thompson Canyon. The canyon is narrow too. Mostly there is just room for the river and the road that run alongside it. But in spots, there is room for cabins, a motel, or a restaurant.

Late in the afternoon that July 31, it began to rain. Local weather stations advised that there could be some local flooding. But otherwise, no one seemed concerned.

However, huge thunderstorms were forming over the upper section of the 78-mile-long canyon. At first, the rain was just a sprinkle. But by 6:30 P.M., the sprinkle had turned into a downpour. In fact, a foot of rain was to fall in the narrow part of the canyon in less than five hours. (Normally, this area receives 2 inches of rain during the whole month of July.)

Meanwhile, people went indoors. They settled into their motel rooms, campers, cabins, or favorite restaurants. No one had any warning of the terror that was about to strike.

Big Thompson River rose rapidly. In some areas, it rose more than 20 feet. It swept down the canyon. There was no escape.

The people in the Covered Wagon Café watched in disbelief as the waters carried cars, trucks, and campers away. Then the café itself was ripped from its foundation and hurled down the mountain. No one inside it survived.

Nothing was left of the Waltonia Motel but bodies and a sodden guest book, found later downstream. The building itself had broken up and disappeared.

The flood turned the two-story hydroelectric plant at the foot of the canyon into a pile of rubble. It left a concrete bridge in shambles. It made chunks of asphalt out of U.S. Highway 34.

The flood killed 139 people that night. It caused $36 million in property damage. It was the worst natural disaster in Colorado history.

Today, Big Thompson Canyon is once again a tourist haven. The scars of the tragedy have faded. But throughout the canyon are warning signs. They show a picture of a man running up a pine-covered hill with waves of water lapping at his ankles. The sign reads: "Climb to safety in case of a flash flood!"

Name ______________________________

Date ______________________________

FLOODS AND TSUNAMIS
REVIEW 1

Get the Facts

1. Why was Big Thompson Canyon full of people the day of the flood?

2. How did the structure of the canyon contribute to the danger of a flash flood?

3. Describe some of the damage done by the flood.

What Do You Think?

4. What warning signs might people in the café have heeded to save themselves? Why do you think they chose to remain in the café?

5. Smaller flash floods took place in Big Thompson Canyon in 1921, 1945, and 1965. Why do you think more people in the area were not aware of the danger?

Teacher's Guide for pp. 96–99

This teacher material covers the following reading selections:

What Is a Flood?
Flash Floods
Seasonal Flooding
Failure of Dams and Levees

Vocabulary (See page 111 for definitions.)

dry wash	generator
flash flood	groundwater
flood	levee
floodplain	reservoir
flood stage	saturated
floodwall	turbine

Answer Key (Review 2, p. 99)

1. (any three) seasonal floods; melting snow; heavy rain; flash floods; a dam or levee breaks; storm surges; tsunamis
2. an area of flatland alongside a river
3. for transportation, food, water, good soil, recreation, beauty
4. Large areas are paved so water cannot soak into the ground; storm drains release large amounts of water all at once.
5. flash floods
6. flood control; to generate electricity; recreation
7. poor materials or engineering; earthquake or avalanche; heavy rains that stress the dam to the breaking point
8. They soak up large amounts of water.

Answers for the following will vary.

9. It may be an attractive site. It may be conveniently located. They may be unaware of the danger. There may not have been a flood for years, so the danger has been forgotten.
10. Yes. Huge amounts of property damage can be done in a flood. Floodplains should be limited to uses such as farming.

Additional Information on Floods, Levees, and Dams

- When a river floods, sediment is deposited on the floodplain. With each flooding, this sediment builds higher. Thus, the floodplain ends up with rich, fertile soil. It becomes good land for farming.
- Large pieces of sediment form a ridgelike deposit that serves as a levee along the side of the river. People build riverbank levees for flood control.
- Some scientists say that flooding is getting increasingly worse as a result of global warming or the greenhouse effect. Heat being trapped in the atmosphere near the earth is warming the earth's climate. With warmer ocean waters, snow and ice are melting faster. Rapid melting is causing the level of the ocean to rise. If the level continues to rise at the current rate, it could be 3 to 5 feet higher 50 years from now. That would mean massive flooding of coastal regions.
- Dams can be used successfully to control floods. However, there are side effects from damming a river. Land downstream no longer receives the nutrients from the floodwaters. Artificial fertilizers must be used instead. Also, water backs up behind the dam, forming a reservoir. People living in the backup area must relocate.

Floods and Tsunamis

What Is a Flood?

Water covers three quarters of the earth's surface. Everything that lives on earth depends on water for life. In fact, two thirds of the human body is made up of water. Both too much water and too little water can be deadly. Too much can cause floods. Too little can cause droughts.

A flood happens when a large, uncontrollable quantity of water covers an area that is normally dry. Floods can cause billions of dollars of damage. And they can kill. In the United States, more people die in floods than in any other type of natural disaster.

Floods have a variety of causes. Seasonal flooding occurs when melting snow or heavy spring rains cause rivers to overflow. Flash floods happen when rain falls so fast that the ground can't soak it up. Coastal flooding comes as a result of storm surges or tsunamis. And floods also occur when dams or levees break.

Flash Floods

Usually, it takes days or weeks for a river to reach flood stage. So people living nearby have time to get ready. This is not the case with flash floods. They strike suddenly with little or no warning. Most flood deaths are caused by flash floods.

Flash floods can be caused by cloudbursts from thunderstorms. They can also be caused by heavy rains accompanying a hurricane. In both cases, the rainfall is so heavy that the ground cannot absorb it. Massive runoff occurs. And the runoff overwhelms rivers and streams. In the desert, dry washes turn into raging streams in minutes. That was what happened in the 1976 Big Thompson River disaster in Colorado.

Seasonal Flooding

An area of relatively flat land alongside a river is called a floodplain. Many floodplains flood every year. Others do so only occasionally. Today, more and more people are building on floodplains. In fact, some cities are built almost entirely on floodplains. Two examples are New Orleans and Cincinnati. Why do so many people chose to build on a floodplain?

Throughout history, people have built homes and cities on the banks of rivers. Rivers can provide transportation, water, and a source of food. Land near rivers is often good for farming. That is because the soil receives valuable nutrients when the river overflows the land. Rivers can offer recreation such as fishing or boating. And usually they offer beautiful scenery as well.

People who build on a floodplain may not be aware of the danger. They may think that flood-control measures like levees or dams will prevent all future floods. If their area has not flooded for years, they may forget it could happen again.

When it rains, the ground normally soaks up the water. Trees and other plants absorb huge amounts of water through their roots. Other water goes deep into the earth, and becomes groundwater.

But when too much rain falls, the ground becomes saturated. It cannot hold any more. The extra water flows downhill. Eventually, it enters a stream or river. The more water that enters, the more the level in the river rises. When the river reaches flood stage, it overflows its banks. This whole process usually only occurs after days of heavy rain.

Human "improvements" often make the flooding worse. In cities, large areas of ground are covered with roads and buildings. Obviously, these areas cannot soak up water. Storm drains are constructed to collect the runoff. Huge amounts of water from these drains pour into streams all at once. That makes flooding worse.

Spring flooding also occurs in areas with large amounts of melting snow. When the weather warms quickly where the snowfall has been heavy, floods can result. Rain falling on the snow can speed the melting and cause flooding.

Failure of Dams and Levees

People build earthen levees along the riverbanks. These levees let the river rise higher without overflowing. People also build concrete floodwalls or dikes.

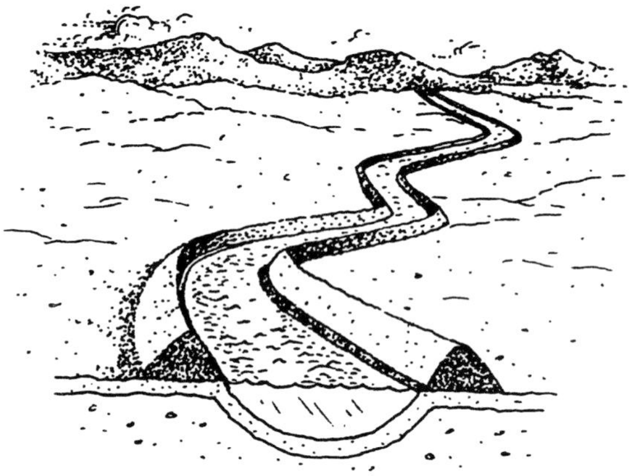

Earthen levee

These structures prevent much flooding. But if they fail because of poor construction, lack of maintenance or just too much water, the result is heavy flooding.

Like levees, some dams are made of earth. When the water behind an earthen dam or levee rises, it creates great pressure on the structure. If the water begins flowing over the top, it may wash away parts of the structure. The pressure of the huge amount of water can cause the structure to collapse, usually all at once. An incredible wall of water then surges out. When the South Fork Dam in Johnstown, Pennsylvania, broke in 1889, thousands of people were drowned by a 40-foot wall of water.

Concrete dams are stronger than earthen dams. They are built across rivers to provide flood control as well as electric power, called hydropower. Upstream from the dam, water

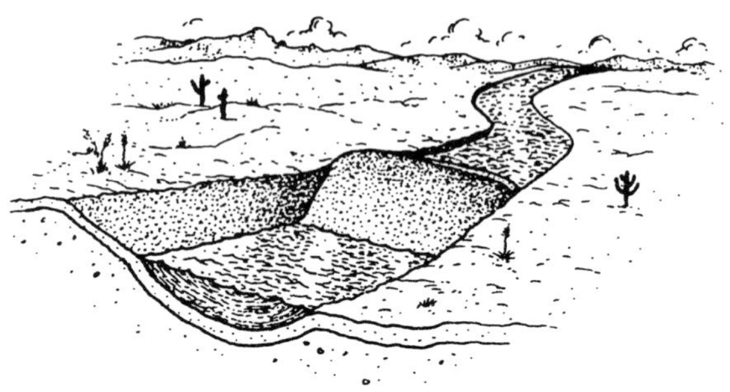

Earthen dam

backs up creating a reservoir. Some of the water is released to run over and turn the turbines. The turbines are connected to generators, which produce electricity.

Water is also released through floodgates. It flows downstream at a gradual, controlled pace. Most of the time the dam works well. But even concrete dams have been known to break, causing massive flash floods.

Dams may fail because of poor design, poor materials, or poor maintenance. They may be located in an area subject to earthquakes or avalanches. In cases of heavy rains, too much water may be released at once from the floodgates. In any case, when a dam fails, a disaster is sure to happen.

Concrete dam

Name ______________________________

Date ______________________________

FLOODS AND TSUNAMIS
REVIEW 2

Get the Facts

1. List three causes of floods.

2. What is a floodplain?

3. Why have people built cities near rivers throughout history?

4. Name a human "improvement" that has worsened flooding.

5. What type of flood causes the most deaths?

6. Name two purposes for building concrete dams.

7. What are some reasons a dam might fail?

8. How do trees and other plants help prevent flooding?

What Do You Think?

9. Many people build homes and businesses in the floodplain. Why do you think someone might choose to do this, even though there is a danger of floods?

10. Some cities have laws that prevent building in a floodplain. Do you think such laws are a good idea or not? Explain your answer.

Teacher's Guide for pp. 101–103

This teacher material covers the following reading selections:

What Is a Tsunami?
What Is a Storm Surge?

Vocabulary (See page 111 for definitions.)

atmospheric pressure	storm surge
amplitude	tsunami
continental shelf	undertow
stationary	wavelength

Answer Key (Review 3, p. 103)

1. undersea earthquakes or volcanic eruptions
2. up to 800 mph
3. It gets much shorter.
4. It gets much taller.
5. Honolulu, Hawaii
6. huge waves that come on shore during a hurricane
7. because the wind must blow in one direction for a period of time to pile up water
8. The atmospheric pressure is low inside the eye, so water is lifted up into it.
9. The land is very low, so there is no place to get away from the water. Also, the population density is great.
10. Galveston, Texas, in 1900

Answers for the following will vary.

11. By the time you can see a tsunami coming, you will not be able to outrun it and get to safety.
12. It is a fertile farming area, and there is always a shortage of food in India and Bangladesh. The people are very poor and have nowhere else to go.

Additional Information on Tsunamis and Storm Surges

- Many people incorrectly call tsunamis and storm surges "tidal waves." Neither of these waves has anything to do with tides.
- "Tsunami" is a Japanese word meaning "harbor wave."
- A tsunami is the result of a sudden uplift or drop of the ocean floor. In an uplift, the leading edge of the tsunami is a wave crest (the high part of a wave). In a drop, the leading edge is a wave trough. When a trough arrives, the ocean recedes from the shore.
- Storm surges are caused by the interaction of wind and low pressure. Strong surface winds push the water, creating waves. These waves enter an enclosed area like a bay. They have no place to go except onto the shore. Hurricanes can have an internal pressure as much as 2.5 inches of mercury lower than the pressure outside the storm. This causes an upward bulge in the water of the sea.

Floods and Tsunamis

What Is a Tsunami?

Tsunamis are massive waves caused by undersea earthquakes or volcanic eruptions. The seafloor lifts up suddenly, displacing huge amounts of water. Tsunamis are sent out in all directions. They travel at speeds of up to 800 miles per hour. They can go thousands of miles without weakening.

In the open ocean, the giant wave may be only a foot or two high. But under the surface, it may be thousands of feet deep. Eventually, this huge water mass reaches the continental shelf. Its wavelength (the distance from the crest of one wave to the crest of the next) may be as much as 100 miles. Its above-water amplitude (height) is only a few feet. But as the wave nears shore, its wavelength shortens and its height grows dramatically.

By this time, the tsunami may be 100 feet high. When the monster wave comes on shore, it quickly floods large areas. And the undertow of the wave, the backward pull under each crest, drags people, houses, and cars out to sea.

The tsunami is not one, but a series of waves which arrive about 10 to 20 minutes apart. The waves may continue to arrive for hours.

The first tsunami warning system was set up in Honolulu, Hawaii, in 1948. This system can pinpoint the origin of earthquakes in the Ring of Fire area. With this information, it is possible to predict the time tsunami waves will arrive at a certain point. This makes it possible to evacuate people from low-lying coastal areas.

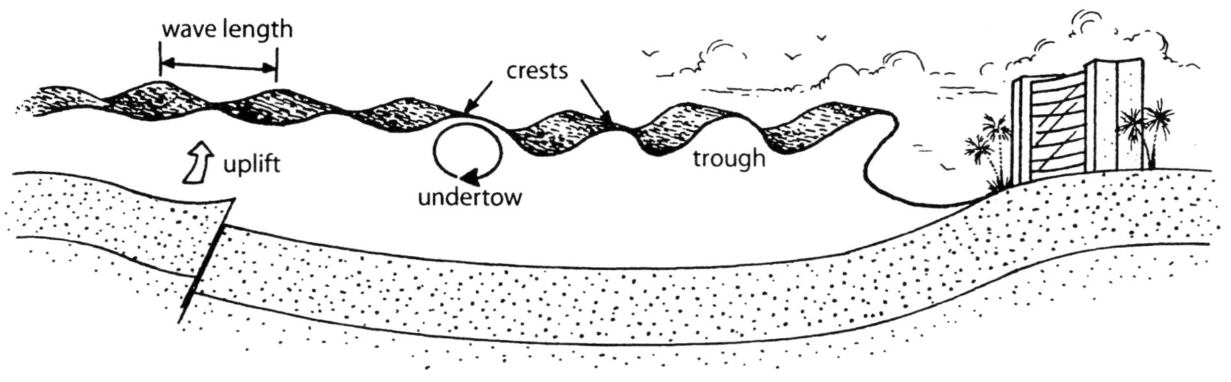

Tsunami

What Is a Storm Surge?

Storms in the ocean such as hurricanes, cyclones, or winter storms produce very strong winds. These winds and the accompanying rain often create ocean waves that are up to 20 feet high. When these waves reach shore, they may travel miles inland. This type of flooding is called a storm surge. In hurricanes, storm surges cause more deaths than strong winds do.

The strength of the wind alone does not cause the storm surge. The wind must blow in one direction for a period of time. That steady push with enough force behind it causes the surge. Thus, a slow-moving or stationary storm is more likely to pile up a storm surge than a fast-moving storm.

The eye of a hurricane or cyclone has very low atmospheric pressure. In the ocean directly below the eye, this low pressure can produce a 3-foot rise in sea level. This may not sound like a lot. But a 3-foot mass that is several miles across is a huge amount of water. In a confined area like a bay, that mass combined with a storm surge and high tide equals a tremendous amount of water coming ashore in a short time.

The Ganges River delta in the Bay of Bengal, between India and Bangladesh, has suffered repeatedly from storm surges. A 1970 cyclone produced a storm surge 15 feet high. That surge drowned 500,000 people. The huge death toll was due in part to the very flat land that offered no escape. It was due also to the population in the area—1400 people per square mile.

The worst storm surge disaster in the United States occurred in Galveston, Texas, in 1900. A hurricane produced a surge 16 feet tall. More than 6000 people drowned. The town was rebuilt with a large seawall to protect it from future surges.

A storm surge caused the flood following Hurricane Camille in 1969. See pages 62–64 for a description of that flooding.

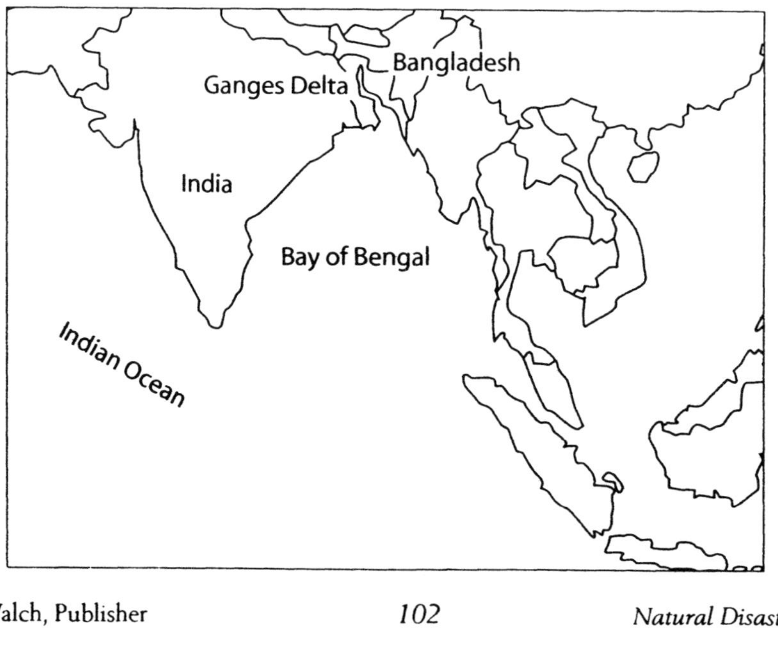

Name ______________________________

Date ______________________________

FLOODS AND TSUNAMIS
REVIEW 3

Get the Facts

1. What causes tsunamis to form?

2. How fast can a tsunami travel?

3. How does the wavelength of a tsunami change as it nears shore?

4. How does the height of the tsunami change as it nears shore?

5. Where is the first tsunami warning center located?

6. What is a storm surge?

7. Why is a slow-moving storm more likely to cause a storm surge than a fast-moving storm?

8. Why is there a rise in the sea level inside the eye of a hurricane?

9. Why is the death toll from storm surges so high in the Ganges River delta?

10. What was the worst storm surge disaster in the United States?

What Do You Think?

11. After an earthquake, some curious people go to the beach to see if they can see a tsunami. Why isn't this advisable?

12. The Ganges River delta suffers regularly from storm surges. Why do you think people continue to live there?

Teacher's Guide for pp. 105–107

This teacher material covers the following reading selections:

How Can You Be Prepared for a Flood?
How Can You Keep Safe in Flooding?
How Can Floods Be Predicted and Prevented?

Vocabulary (See page 111 for definitions.)

diversion channel
flood insurance
flood watch
flood warning
meteorologist
wetlands

Answer Key (Review 4, p. 107)

1. Ask your local Red Cross chapter or your city government.
2. A flood watch indicates that a flood is possible. A flood warning indicates flooding will probably occur.
3. Head for high ground and stay out of the water.
4. (any three) Even shallow moving water can sweep you off your feet. Downed power lines may charge the water. Floodwater can hide debris, animals, and broken glass. Water may suddenly become deeper than it looks. Cars may stall. The water may be contaminated by sewage.
5. It is likely to be contaminated by sewage.
6. computers, weather satellites
7. to divert flood waters away from a city into another body of water
8. They soak up huge quantities of water and allow it to seep slowly away.

Answers for the following will vary.

9. **Pro:**

   The location would be convenient to town.
   The homeowners could buy flood insurance to protect them.
   It hasn't flooded in ___ years and may not flood again.
   We could build a levee or dam for protection.
   The city could gain much property tax money from these homes.
   It will help the city's economy to build businesses here.

   Con:

   It will destroy wetlands, causing more flooding and the destruction of wildlife habitat.
   People can build somewhere else.
   It may not have flooded in years, but it will again.
10. It may be a convenient location. It may be cheaper. It may have good soil.

Additional Information on Flood Damage and Deaths

- Floodwaters are very destructive to property. When they enter homes, they bring sewage and mud with them. They destroy furniture, appliances, carpets, and walls. Most distressing can be the loss of family heirlooms or photographs.
- Most flood deaths occur when people are hit by objects carried along by floodwaters.

How Can You Be Prepared for a Flood?

Find out what the risk of flooding is where you live. Your local chapter of the Red Cross or your city government can tell you. If you live in a flood-prone area, make sure your homeowner's or renter's insurance covers flood damage.

If it has been raining hard, pay attention to radio, television, and newspaper announcements of flood danger. If a flood watch is issued, a flood is *possible*. If a flood warning is issued, flooding will *probably* occur. If you hear a flood warning for your area, lock up and head for higher ground.

A flash flood watch means flash flooding *may* occur. A flash flood warning means flash flooding *is* occurring or *will soon occur*. If you hear such a warning, evacuate right away. You may have very little time to reach higher ground.

How Can You Keep Safe in Flooding?

According to the American Red Cross, the most important rule for staying safe in a flood is this: Head for high ground and stay out of the water. In a flood, the most dangerous thing you can do is to walk, swim, or drive through floodwater. Even very shallow moving waters can sweep you away. Downed power lines may cause electrocution. Floodwaters may hide broken glass, snakes, or debris. Cars may become stalled in water that is deeper than it looks. If your car stalls in rising water, get out and run for safety.

Water is everywhere after a flood. But don't drink any of it. Flooding usually contaminates the drinking water supply. It causes waste disposal systems to break down. So drink bottled water and boil all tap water before you use it. Throw out any food that has come in contact with the water.

Listen to the radio or TV for news about what to do and where to go for help. The Red Cross and the local government will provide help in emergencies.

How Can Floods Be Predicted and Prevented?

Meteorologists are scientists who study the weather. At the National Weather Service, they use weather satellites to gather information. Computers help analyze that data. Then the meteorologists issue predictions of severe thunderstorms, hurricanes, and flash floods.

The National Weather Service issues warnings when a river is in danger of flooding. The National Hurricane Center in Miami, Florida, tracks hurricanes. All this information is vital. It gives people time to prepare and save themselves and perhaps treasured family items as well.

It is not possible to control the weather that causes the floods. So people build protective barriers—dams, levees, dikes, flood walls, and seawalls.

People also build channels to divert some of the floodwater. They build these channels around areas of a river that often flood. The channels direct the water into other bodies of water. In New Orleans, a diversion channel sends some of the water from the Mississippi River into Lake Pontchartrain.

Preservation of wetland areas is important to flood control. Wetlands such as marshes and swamps are important natural flood protection. These wetlands soak up huge amounts of water. And they drain it off slowly. When people fill in wetlands to build houses, roads, or businesses, they destroy natural flood protection. Thus, flooding may become a problem in the area.

The best protection against flood damage is to *not* build a home on a floodplain or in low-lying areas such as beaches or marshes. In some states laws have been passed that forbid such building. But it is difficult to keep people from living where they want to. So each year, many people suffer damage from floods.

The Mississippi River has had more engineering work done on it than any other river. Levees, diversion channels, concrete flood walls, and reservoirs have tried to tame the 2340-mile-long river with some success. Floods continue to occur. But they are much less severe than they would have been without human intervention.

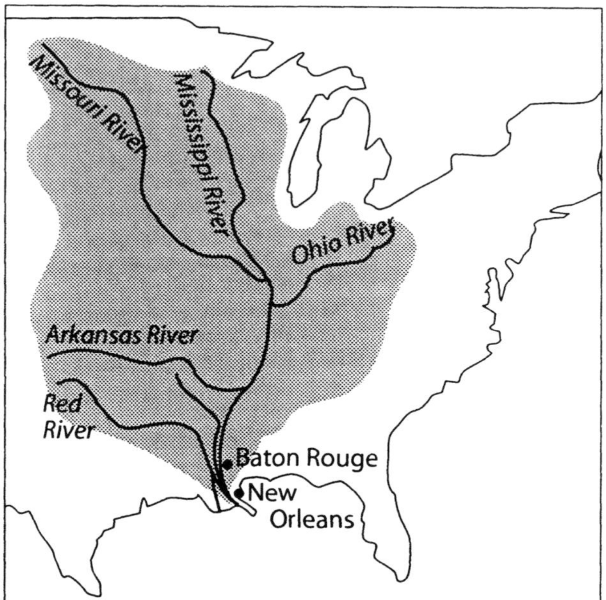

The basin of the Mississippi River

Name ______________________________

Date ______________________________

Floods and Tsunamis
Review 4

Get the Facts

1. How can you find out if the area you live in is prone to floods?

2. Explain the difference between a flood watch and a flood warning.

3. What is the most important rule for staying safe in a flood according to the American Red Cross?

4. Name three reasons floodwater may be more dangerous than it looks.

5. Why shouldn't you drink tap water after flooding has occurred?

6. Name two tools meteorologists use to predict the weather.

7. What is the purpose of a diversion channel?

8. Why are wetlands important to flood control?

What Do You Think?

9. Imagine that your city council is debating whether to allow development in a wetlands area next to your town. What do you think would be some arguments for and against the development?

10. Why do you think people might buy or build a home in an area that has flooded before?

Floods and Tsunamis

Additional Information

Worst Floods in World History

Year	Location	Cause	Deaths
1412 B.C.	Crete	Tsunami caused by volcano	unknown
A.D. 365	Egypt	Tsunami caused by earthquake	50,000
1099	England/Holland	Storm surges and high tides	100,000
1287	Holland	Seawall collapses	50,000
1642	China	River dikes destroyed in war	300,000
1692	Jamaica	Tsunami caused by earthquake	1600
1755	Portugal	Tsunami caused by earthquake	50,000
1824	Russia	Ice jam in river	10,000
1876	India	Cyclone	200,000
1881	India	Cyclone	100,000
1883	Java	Tsunami from Krakatoa eruption	36,000
1887	China	Yellow River floods	900,000
1896	Japan	Tsunami	27,000
1931	China	Yellow River floods	145,000
1938	China	Dikes blown up in war	1,000,000
1942	India	Cyclone	40,000
1963	Bangladesh	Cyclone	22,000
1970	Bangladesh	Cyclone	250,000
1995	China	Yangtze River floods	1200

Worst Floods in the United States

Year	Location	Causes	Deaths
1861	California	Sacramento River flood	700
1874	Massachusetts	Dam failure	144
1889	Johnstown, PA	Dam failure	2000
1900	Galveston, TX	Storm surge/hurricane	6000
1903	Kansas City, MO	Heavy rains	200
1913	Ohio, Indiana	Ohio River flood	732
1928	California	Dam failure Santa Paula	450
1972	South Dakota	Dam failure	242
1972	West Virginia	Dam failure	1125
1976	Colorado	Flash flood/Big Thompson	139
1990	Texas, Oklahoma	River floods	23
1993	Midwest	Mississippi River floods	60

(continued)

Additional Information *(continued)*

- The National Oceanic and Atmospheric Administration (NOAA) includes the National Weather Service, National Ocean Survey, National Environmental Satellite Service, Environmental Research Laboratories, Environmental Data Service, and National Marine Fisheries Service. It publishes data on tides and currents.
- Christopher Columbus experienced a tsunami on August 14, 1498, while sailing between Trinidad and Venezuela.
- Venice, Italy, is world-famous for its architectural and cultural heritage. The city is said to have been settled by people fleeing from Attila the Hun. These fugitives picked a poor spot on which to build a city. The city is a collection of 118 islands in the lagoon of Venice. It is built on piles sunk deep into the mud. It is linked to the mainland by a causeway. At present, the Adriatic Sea is rising and Venice is sinking. To travel Venice's streets, you need a boat.
- Engineers have long looked for ways to use the power of the tides. Tidal power could be the perfect energy source. It is inexhaustible, nonpolluting, and always dependable. The first large-scale tidal power plant was built in France. That plant in the estuary of La Rance has been operating since 1966. Since then, plants have been built in other countries. These plants have one drawback. They do not produce power uniformly. They are most productive when the tides are highest.
- Holland has fought the sea's flooding for centuries. The small country, now known as the Netherlands, has a long history of disasters from storm surges off the North Sea. In 1099, 100,000 were killed. In 1287, 50,000 drowned when a seawall collapsed. Other disasters occurred in 1421, 1570, 1825, 1894, and 1916. Then in 1953, dikes were breached in a hundred places. One sixth of the country was flooded. Nearly 1800 people died. That disaster was the impetus for new feats of engineering to keep back the sea.
- The Yellow River (Hwang He) in China is the most flood-prone river in the world. It is called China's Sorrow because of the many deaths it has caused. It has flooded 1500 times in recorded history. In 1887, it overflowed the 70-foot-high levees in Honan Province. Some 900,000 people drowned. Yet the river remains one of the most populated areas of China. It offers riverbank residents food, transportation, water, and soil.
- Floods are the world's most widespread and damaging type of natural disaster.

Activities for Students

1. Look for more information on the ecology of wetlands. Why are we losing vast amounts of wetlands each year? Why are wetlands important? What can be done to save the wetlands?
2. Find out which areas in your county are subject to flooding. Report on any recent flooding and the damage it caused.
3. Report on a famous flood.
4. Read about a recent hurricane. Report on the damage done by storm surge in this hurricane. What caused the most damage and deaths, high winds or storm surge?
5. Invite an official from the local Red Cross or Emergency Preparedness Department to speak to your class. Ask the official to discuss disasters that have occurred in the area and ways to be prepared for future disasters.
6. Report on Atlantis, the mysterious land, that may have disappeared in a tsunami 3400 years ago.
7. Find out more about one of these tsunami disasters:
    - the destruction of Alexandria, Egypt, in A.D. 365
    - the devastation of Port Royal, Jamaica, in 1692
    - the destruction of Lisbon, Portugal, in 1755
    - the destruction of much of Sumatra and Java by the eruption and tsunami caused by Krakatoa in 1883
    - the Hilo, Hawaii, tsunami of 1946
    - the tsunami caused by the Alaskan earthquake in 1964
8. Read more about the Ganges River delta. This area lies at the head of the Bay of Bengal in India and Bangladesh. It has been devastated many times by storm surges. The result has been huge numbers of deaths.
9. In 1953, a storm in the North Sea created havoc in the Netherlands and England. Read more about this storm.
10. Find out more about the system of dikes in the Netherlands.
11. Read about the destruction of Galveston, Texas, by a storm surge in 1900.
12. Find out more about the National Oceanic and Atmospheric Administration (NOAA).
13. Read more about tidal power, creating electricity by harnessing the power of waves.
14. Read about the story of Venice, Italy.

Floods and Tsunamis

Vocabulary

amplitude	height of a wave
atmospheric pressure	force exerted by the atmosphere
continental shelf	shallow slope around a continent, outside of which the ocean floor drops off sharply
diversion channel	channel constructed for draining off flood waters
dry wash	gully formed by water erosion, which only has water during flash floods
flash flood	quick flooding of an area
flood	covering by water of land that is usually dry
flood insurance	insurance that covers losses due to flooding
floodplain	relatively flat area that runs alongside a river
flood stage	stage at which a river begins to overflow its banks
floodwall	concrete wall alongside a river meant as flood control
flood warning	announcement that a flood is likely to occur in an area
flood watch	announcement that there is a possibility of flooding
generator	machine that converts water energy to electrical energy
groundwater	water lying beneath the earth's surface
levee	barrier made of earth that runs along a river to keep it from flooding
meteorologist	scientist who studies weather
reservoir	lake constructed behind a dam, which holds water for drinking or making electrical power
saturated	holding as much water as it can
stationary	not moving
storm surge	huge waves and rise in sea level that bring water onto shore as a result of a hurricane
tsunami	huge waves caused by undersea quakes or volcanic activity
turbine	engine that turns because of the pressure of water or gas
undertow	current that moves backward out to sea
wavelength	length between the crests of two waves
wetlands	land with wet and spongy soil such as in a marsh or bog

When It's Hot and Dry

Teacher's Guide for pp. 115–116

This teacher material covers the following reading selection:
The Deadliest Fire in North America

Vocabulary (See page 128 for definitions.)

bogs   slash

Answer Key (Review 1, p. 116)

1. The country was suffering from a prolonged drought.
2. Towering forests surrounded the town. Transportation was readily available for the wood products.
3. Fierce winds blew the fire.
4. They left piles of debris about to fuel the fire.

Answers for the following will vary.

5. Chicago was larger and better known. Chicago had major newspapers, which naturally concentrated on their city's fire. Peshtigo was cut off from the world and was little known. Property damage was incredible in Chicago. Much of the city burned.
6. The danger did not appear to be imminent. Small fires had been burning for months.

Additional Information on the Peshtigo and Chicago Fires

- The Peshtigo fire stimulated the creation of the National Forest System.
- In the Peshtigo fire, most people died from asphyxiation. The rapidly moving fire used up all the oxygen, leaving none to breathe.
- The 1871 drought had dried out the wooden buildings and wooden sidewalks in Chicago. Everything was so dry that most insurance companies canceled fire protection policies.
- Legend has it that the Chicago fire was started by Mrs. O'Leary's cow. The cow wanted to be milked, so it began to kick. It kicked over a burning kerosene lantern, which ignited the hay. It was proved that the fire started in the O'Leary barn. But Mrs. O'Leary denied that her cow started it.
- When the Chicago fire began, people tried to sound the alarm. But the new fire alarm boxes did not work. The fire spread quickly, aided by fierce winds. The city's waterworks were destroyed. So there was no water to put out the blaze. In the end, 300 people were killed and 100,000 were left homeless. The heart of the city—government buildings, hotels, factories, businesses—was gone. On October 10, the rain began. Four hours later, the fire was out.

The Deadliest Fire in North America

In the year 1871, the entire continent of North America suffered a severe drought. It was the worst drought anyone could remember. There had been little snow that winter. When spring came, there was very little water in the rivers.

Then summer arrived, hot and dry. Almost no rain fell anywhere. By fall, the forests of Wisconsin were as dry as kindling.

Peshtigo, a lumbering community in northern Wisconsin, was a boomtown then. The community was so prosperous that people said it would soon be larger than Chicago.

Peshtigo was an ideal spot for lumbering. Thick forest pressed in all around the town. Transportation was readily available. Wood products were shipped from the harbor at Green Bay to Chicago. From there, they were sent all over the country.

Meanwhile, crews were building a railroad to link Peshtigo with the outside world. They worked quickly to lay track, cutting down trees and brush. They left behind piles of debris.

Lumbering operations also were making inroads into the great forest. And, as a result, more debris, or *slash* was left piled up.

By summer's end, small fires were breaking out in scattered areas of the forest. Under the surface of the bogs, fires were widespread. Here and there they would break through to the surface.

By October there was so much smoke in the air that ships on Green Bay had to navigate by compass and use their foghorns. Telegraph lines burned. Fire swept down the road leading into town. Peshtigo was cut off from the outside world.

Some people were concerned. Others thought that the autumn rains would arrive in time to prevent any real danger. People began to stream into town from the burning countryside. The scattered fires had begun to merge.

On Sunday evening, October 8, a rumbling shook the forest. An ominous glow lighted the sky. The fire had arrived in Peshtigo. Burning branches and embers fell on the town from the sky. Hurricane-strength winds began to blow, sweeping flames across the town.

Most of the people were simply engulfed in flames. Some were able to make it to the Peshtigo River. There, they went under water to escape the flames. But, in the end, many were burned when they came up for air.

Within minutes the fire destroyed the entire town, killing more than 1500 people. It destroyed some 4 million acres of forest. The next day, it began to rain. The fire was extinguished. But not before it had taken more lives than any other North American fire.

Yet, few people have ever heard of the Peshtigo catastrophe. That is because of a coincidence. On the night of the Peshtigo fire, October 8, 1871, the city of Chicago was set ablaze. This famous fire, said to have started when Mrs. O'Leary's cow kicked over a lantern, killed 300 people and destroyed much of Chicago's downtown area.

Name ______________________________

Date ______________________________

Fires
Review 1

Get the Facts

1. Why was the forest surrounding Peshtigo so dry?

2. Why was Peshtigo an ideal spot for lumbering?

3. Why did the fire sweep through town so quickly?

4. In what way did the lumbering and railroad crews help the fire burn?

What Do You Think?

5. Why do you think the Peshtigo fire got so little attention in the press compared to the Chicago fire?

6. Why do you think people in Peshtigo were not more concerned about the fires in the surrounding forest?

Teacher's Guide for pp. 118–120

This teacher material covers the following reading selections:

What Causes Forest Fires?
What Are the Kinds of Forest Fires?
What Is the Nature of Forest Fires?

Vocabulary (See page 128 for definitions.)

arsonist	incendiary fire
crown fire	oxidation
firestorm	pyromaniac
ground fire	surface fire

Answer Key (Review 2, p. 120)

1. drought, wind, and heat
2. people
3. a fire set by an arsonist
4. It may break through to the surface and set surface material on fire.
5. a crown fire, or firestorm
6. crown fire that gets out of control due to high winds generated by the fire
7. fuel, heat, and oxygen
8. the plants there have too high a water content
9. uphill
10. Some plants are drier; others have a higher water content.

Answers for the following will vary.

11. Extinguish campfires carefully. Shovel dirt on the fire when leaving a campsite. Use ashtrays. Do not drop cigarette butts outdoors. Be careful when choosing a site for a campfire. Don't build too large a fire. Don't leave the fire unattended.
12. Firebreaks could be constructed around housing developments. Fire-resistant shrubs could be planted in these areas. No homes should be built until adequate water is available for firefighting. Building could be restricted in particularly dangerous areas.

Additional Information on Forest and Brushfires

- Materials must reach their kindling temperature before they can burn. The kindling temperature of different types of wood varies with moisture level, density, and chemical makeup. Wood containing large amounts of resins or oils produce flammable gases as they burn.
- The kindling temperature of most wood is 600°F. But rotting or very dry wood may have a kindling temperature as low as 480°F.
- Some ground fires are started by spontaneous combustion. As a compacted mass of material decays, heat is generated and stays within the mass. Eventually it may reach kindling temperature.
- Brushfires may move at speeds of 6 mph, or even faster in the case of grass fires. These rapidly moving fires do not usually damage mature trees, unless a lot of litter has accumulated and the fire becomes intense. Aided by winds, crown fires may move through the trees at 20 mph or faster.

Fires

What Causes Forest Fires?

We all know it's easier to start a fire with dry wood than with wet wood. In the same way, a forest fire is most likely to start when the woods are very dry. Drought, wind, and heat all cause trees and brush to dry out and become prone to catching fire.

How do forest fires start? Lightning starts perhaps 10 percent of the forest fires in the western United States. Humans start the other 90 percent. Some of the 90 percent are sparked by lumbering operations. Others are caused by careless campers or hikers. Discarded cigarettes start some fires. And fully one third of all fires are deliberately set by arsonists. Another name for these fires is incendiary fires.

Why would anyone start a fire? Some people hope to collect insurance money. Others want revenge on someone they don't like. Still others are pyromaniacs—people with an uncontrollable urge to set fires.

What Are the Kinds of Forest Fires?

Ground fires burn under the surface of the ground. They smolder in thick masses of organic material like roots, leaves, and litter. They may burn for months, spreading slowly. They damage tree roots and living matter in the soil. Sometimes they break through and start a surface fire.

Surface fires burn grass, shrubs and other material above ground. Surface fires often move very quickly. They rarely harm mature trees. And they are usually easy to control.

Crown fires have spread to the treetops (crowns). In crown fires, all levels of the forest are burning. These fires are hard to control. They often produce their own winds, which spread burning material farther.

When they get out of control, they move very quickly. At this point, they become firestorms. It was a firestorm that destroyed the town of Peshtigo, Wisconsin, in 1871.

What Is the Nature of Forest Fires?

For centuries it was believed that fire, like water, was a substance. In the 1700's, scientists discovered that fire is a process called combustion. In fact, fire is an oxidation process, the same process that forms rust. Of course fire is a much faster and more violent form of oxidation.

Three elements are necessary for fire to burn. They are fuel, heat, and oxygen. By eliminating any one of these, the fire can be put out.

The behavior of a forest fire depends on three factors: the amount and type of fuel available, the weather, and the terrain where the fire is burning.

Fuel

The size and kind of plants available affect how a fire will spread. Taller plants enable the fire to reach greater heights more easily. Some types of plants burn more readily than others. The chaparral plants that cover much of southern California, for example, have been said to be "born to burn." The tiniest spark sets them ablaze. Dry grasses also burn very easily. Other plants are more fire resistant.

Weather

Forest fires are not apt to occur in a rain forest. Plants in tropical areas are not likely to catch fire since they have a high water content. And they may have water on their surface as well.

On the other hand, plants in areas with very low humidity can quickly become dry. Once dry, they can ignite easily.

Wind spreads fires. It also feeds the fire more oxygen, which makes the flames burn hotter.

Terrain

On flat ground, the spread of a fire is determined by the type of fuel being burned and the amount of wind present. In hilly areas, this becomes more complicated.

Since hot air rises, fire naturally tends to run uphill. Downhill, fire travels more slowly. For this reason, lightning-caused fires often can be controlled. They usually start on a mountain peak. There, they have little fuel, since vegetation is sparse. And they are forced to run downhill.

Most mountainous areas consist of many hills and valleys running different ways. So it is not often clear which path the fire will take, or how fast it will take it.

Name ______________________________

Date ______________________________

FIRES
REVIEW 2

Get the Facts

1. What three factors cause a forest to become very dry?

2. What is the cause of 90 percent of forest fires?

3. What is an incendiary fire?

4. How can a ground fire become a surface fire?

5. Which type of fire is the hardest to control?

6. What is a firestorm?

7. What are the three elements needed for fire to burn?

8. Why do forest fires not normally occur in a rain forest?

9. In which direction do fires move when they encounter a hill?

10. Why do some plants burn more easily than others?

What Do You Think?

11. What do you think are some ways people can avoid accidentally setting a fire when camping or hiking?

12. The dry chaparral-covered hills of southern California are prone to frequent brushfires. These fires destroy many expensive homes. What do you think might be done to reduce fire damage in the area?

Teacher's Guide for pp. 122–124

This teacher material covers the following reading selections:

How Are Forest Fires Fought?
What Are the Effects of Forest Fires?
How Can Forest Fires Be Prevented?

Vocabulary (See page 128 for definitions.)

aerial drop	firebreakers	managed fire
backfire	fire species	smoke jumpers

Answer Key (Review 3, p. 124)

1. U.S. Forest Service
2. to spot fires while they are small and easier to put out
3. to stop fire by burning the fuel in its path
4. (any two) throwing sand or dirt on the fire; cutting down branches or brush; directing water on the fire; constructing firebreaks
5. a cleared space at least 10 feet wide and as long as needed
6. in anticipation of where a fire may go
7. So much litter may build up that when there is a fire it will be a much larger fire.
8. The seeds require heat to break open their hard shells.
9. Nutrients contained in plants that are burned are released into the soil. New plants take them up through their roots.
10. to clear areas of underbrush to avoid a larger fire later

Answers for the following will vary.

11. Arsonists start fires with the idea of causing as much damage as possible. They may start fires in weather conditions conducive to the spread of flames. They may start them in places where damage will be greatest.
12. They may be afraid the fire will get out of control. They have the idea that fire in a forest is always bad. They are concerned about the air pollution caused by smoke.

Additional Information on Fires and Fire Fighting

- People clear firebreaks in anticipation of future fires. They build them at right angles to the prevailing winds. They keep them free of brush and litter. In a further step called a fuelbreak, people grow fire-resistant plants in the firebreak. A fuelbreak lessens maintenance and soil erosion.
- From aircraft, officials drop water or fire retardants on a fire. Fire retardants are thick liquid chemicals that coat plants to increase resistance to heat or flames.
- Firefighters use four-wheel-drive vehicles to carry water and pumps out to remote areas.
- **Early Humans and Fire**

  Our earliest ancestors may have first learned about fire from observing volcanoes and lightning-blazes. At some point, people learned how to keep fire going by feeding it fuel and protecting it from wind and rain. As they moved about, they carried coals with them so they could start new fires.

  Later, people learned to start fires by striking the right kinds of rock (flint) together to make sparks. Others did the same by rubbing sticks together.

  In 1827, John Walker of England invented the first matches. He called them friction lights. Later, phosphorus matches were invented. They were unsafe. They ignited when rubbed against anything and sometimes ignited spontaneously.

  The safety matches of today were invented in Sweden in 1855. They must be struck on a special surface on the cover.

How Are Forest Fires Fought?

Several organizations are responsible for fighting forest fires. The U.S. Forest Service fights fires in national forests. Each state has a forest service or natural resources unit that protects state forests. In addition, local fire fighters respond to forest fires in their area.

The most effective way to fight fires is to spot them quickly and attack them while they are small. Officials watch for fires from lookout towers, which have been set up in fire-prone areas. They radio the exact location of the fire to fire-fighting crews. The crews rush in to put out the fire while it is still small.

Fires crews often use lumbering roads or other trails to reach a forest fire.

Other times, the forest fire is in a more remote region. Then smoke jumpers parachute into an area close to the fire. They wear protective clothing. They carry tools needed to slash brush and create firebreaks. They also carry a small aluminized shelter for use as emergency protection from the fire.

Fire fighting is a dangerous and difficult job. It is not unlike military combat. Some techniques used to fight fires include the following:

Lighting Backfires

Firefighters set backfires to burn the fuel in the path of an oncoming fire. Then when the fire arrives, it has no fuel to burn. Obviously, firefighters only light backfires when there is very little wind.

Aerial Drops

From aircraft, fire officials release water or fire-retardant chemicals to put out the fire. If a fire is very large, this may be the only way to stop it.

Hand Attack

Ground crews throw dirt or sand on the fire. They cut down branches and brush to take away fuel from the fire. They bring water to the fire in tanks, trucks, and pumps. Then they use hoses to spray water on the fire.

Firebreaks

Firefighters clear firebreaks in front of the fire to rob it of fuel. They clear a space at least 10 feet wide and as long as needed. If there is time, they strip the area right down to the soil. They do this by hand or with bulldozers or fire plows.

What Are the Effects of Forest Fires?

Clearly, a large, raging fire is a highly destructive force. Fires take lives, both human and animal. And fires destroy huge amounts of timber. Timber damaged by fire becomes a prime target for insects and rot.

Yet not every forest fire is undesirable. Surface fires get rid of the leaf and branch litter on the forest floor. If this material were left to pile up for years, it would become a huge supply of fuel for a big fire. And when that well-fed fire occurs, it turns into a very large, intense one instead of a simple surface fire.

Fire plays a key role in the survival of some plant communities. These plants are called fire species. The Douglas fir predominates in northwestern forests. It seeds well in areas left open by fires. The ponderosa pine requires periodic fires to burn away the thousands of seedlings that spring up under it. Unless this undergrowth is cleared away, the ponderosa will be replaced by these other trees.

The aspen thrives after a fire by sending out long lateral roots. The roots sprout into trees, which grow into big stands. A group of shrubs called *Ceanothus* requires fire to open the hard coat on its seeds. After a fire, the shrubs come up everywhere. They serve as winter food for many animals, such as deer and elk.

Fires help provide forest animals with extra nutrition. When plants burn, the nutrients they contain are released into the soil. Plants that grow right after a fire soak up these nutrients. Animals that eat these new plants get the extra benefit.

How Can Forest Fires Be Prevented?

The most important part of fire prevention is to make the public aware of fire hazards. To this end, Smokey the Bear was chosen as the symbol of the fire prevention campaign in 1945. Smokey's motto is: "Only YOU can prevent forest fires."

A second part of fire prevention is to reduce the amount of fuel available to burn. This is done by cutting down and removing dead trees and by removing debris left from logging. Sometimes forest rangers set small, controlled fires in areas that need to be cleared of undergrowth.

A third part of fire prevention is to clear firebreaks in strategic areas. Firebreaks are strips of land that have been stripped of flammable materials. Once cleared, they must be maintained to be effective. Firebreaks help contain a fire if one starts. But there is debate about their effectiveness.

Name ______________________________

Date ______________________________

FIRES
REVIEW 3

Get the Facts

1. What organization is responsible for fighting fires in national forests?

2. Why are lookout towers so important in fire fighting?

3. What is the purpose of a backfire?

4. Name two ways firefighters attack a fire by hand.

5. What is a firebreak?

6. Why are firebreaks sometimes cleared years ahead of a fire?

7. If surface fires are eliminated from a forest, what undesirable event may take place later on?

8. Why are surface fires necessary for the growth of *Ceanothus* shrubs?

9. How does fire increase the nutrient level in plants that grow after the burning?

10. What is the purpose of managed fires sometimes set by rangers?

What Do You Think?

11. Why do you think that incendiary fires are usually the most difficult fires to control?

12. Many people object to planned fires in national parks, despite benefits to the ecology there. Why do you think these people might advocate a fire-free policy?

Additional Information on Great Fires

Catastrophic Forest Fires

- The Peshtigo, Wisconsin, fire in 1871 (see pp. 115).
- A huge fire in 1865 in Silverton, Oregon, destroyed a million acres of timber.
- A forest fire in 1871 in Michigan killed 138 people.
- In 1957, a fire in Alaska burned 5 million acres.
- The Tillamook, Oregon, fire of 1933 was started by sparks caused when a log was dragged over a steel cable. The forest was so dry, the flames could not be stopped.
- The Sundance Mountain fire took place in 1967 in a 150,000-square-mile area in the states of Oregon, Idaho, Montana, and Washington.
- In 1987, northern California and southwestern Oregon suffered numerous fires that burned almost a million acres.
- In 1991, a wildfire in Oakland, California, destroyed much property and caused 24 deaths.
- In 1988, well over half of Yellowstone National Park was burned. This was the worst fire ever to occur in a national park. A great debate raged over whether to put out the fires or let them burn as long as they did not threaten human habitation. Until the 1970's, the policy had been to put all surface fires out. The big question, still unresolved, was whether this policy, by allowing so much fuel to build up, had in effect caused the huge fire of 1988. Scientists remain divided on this issue.

Catastrophic Fires in Cities

- The Great Chicago Fire in 1871 (see p. 115).
- A major earthquake on April 18, 1906 led to the San Francisco Fire. The quake downed power lines and upset stoves and lamps, which started numerous fires. At the same time, the quake broke all the water mains in the city. About 75 percent of the city was destroyed. Some 300 people died and 300,000 were left homeless.
- The Great Fire of London in 1666 destroyed most of the city. A new, improved city emerged. Buildings were made of brick rather than wood. A new city water system, and wider streets were built. Two other benefits came as a result of the fire. Fire insurance companies were organized to protect people from losses by fire. And the first fire departments were set up by the insurance companies.
- The most disastrous fire ever to take place within one building was at The Theater in Canton, China, in 1845. A total of 1670 people were trapped in the theater and died.
- In 1944, 163 people were trapped and killed in a flaming circus tent as the Ringling Brothers, Barnum and Bailey Circus performed in Hartford, Connecticut.
- In 1942, 492 people died in a fire at the Coconut Grove nightclub in Boston. The club was very overcrowded. Some exit doors were locked. This incident led to new fire codes and fire inspection policies being adopted in buildings around the country.

(continued)

Additional Information *(continued)*

- On January 27, 1967, three *Apollo* astronauts were killed in a fire on the launching pad in Cape Canaveral, Florida.
- In May 1937, the German airship *Hindenburg* was preparing to land at Lakehurst, New Jersey. Inside its metal shell were giant bags filled with hydrogen gas. This gas was highly flammable. Some 75 feet off the ground, the airship burst into flames. Some people lived by jumping out. A total of 41 were killed.
- Fire has often been used as a weapon of war. In 1812, the British burned Washington, D.C. General William Tecumseh Sherman burned cities across the South during the Civil War. Fire is used as a weapon of war today as well. Flamethrowers have been used against tanks and machine gun locations. Incendiary bombs ignite when they hit the ground. And nuclear bombs result in immense fires.
- A 1993 fire in a cult compound in Waco, Texas, killed 72 people.

Activities for Students

1. Find out how the slash-and-burn technique is used to clear tropical forests.
2. Write a poem or essay contrasting the pleasant aspect of fire (the warmth of fire in a fireplace, candlelight at dinner, bonfires at college, etc.) with the unpleasant side (fire in war, a firestorm, an arsonist, etc.).
3. Read the story of Smokey the Bear.
4. Invite a member of the Forest Service to speak to your class about fire prevention and firefighting.
5. Write a report on one of the catastrophic fires listed on pages 125–126.
6. Find out the beneficial effects surface fires have on one or more of the following species:

ponderosa pine	larch
white pine	Douglas fir
lodgepole pine	aspen

7. Write a report on modern fire-fighting techniques used in forest fires.
8. Find out about the fire in A.D. 64 that destroyed Rome. What part was Emperor Nero said to have played in this fire?
9. Read more about the Yellowstone Park fire of 1988. Opinions were divided on whether to let the fire burn or to put it out. What were the arguments on each side?
10. Read about the October 1991 fire near Oakland, California, that burned 3000 homes and killed 24 people.
11. Read about the Tillamook fire in Oregon, which burned for 18 years between 1933 and 1951.
12. Find out whether there have been any disastrous fires in the area where you live. Report to the class about them.

Fires Vocabulary

aerial drop	dropping of water or fire-retardant chemicals on a fire from a plane or helicopter
arsonist	person who sets fires deliberately
backfire	fire set ahead of the main fire with the idea of burning the fuel before the main fire gets there
bog	area of ground made up of decayed plants that stays wet and soggy
crown fire	fire that reaches and travels through the tops of trees
firebreak	strip of land from which brush and trees have been removed, to stop the spread of a wildfire
fire species	plants that need fire as part of their natural cycle
firestorm	violent fire that develops its own fierce wind system
ground fire	fire that occurs beneath the ground, often in peat or bogs
incendiary fire	fire set on purpose; arson
managed fire	fire set by officials such as the U.S. Forest Service for the purpose of burning away undergrowth to prevent a larger fire later on
oxidation	process of combining with oxygen; in the case of rusting, this is a slow process, and in the case of fire, a quick one
pyromaniac	person with an uncontrollable urge to set fires
slash	debris resulting from the felling of trees in lumbering operations
smoke jumpers	firefighters who parachute close to a forest fire
surface fire	fire that travels along the surface of the ground, burning underbrush and not reaching the tops of trees

Drought

Teacher's Guide for pp. 130–132

This teacher material covers the following reading selection:
The American Dust Bowl

Vocabulary (See page 144 for definitions.)

black blizzard
drought
Dust Bowl
Great Depression
Great Plains
gullies
price supports
sodbusters
soil erosion
subsidies
windbreak

Answer Key (Review 1, p. 132)

1. They had to work hard to break up the sod so they could plant seeds.
2. farmers plowing up the grass; livestock overgrazing
3. Farmers were urged to grow as much wheat as possible; prices were high; the rainfall was ample.
4. Severe droughts occurred.
5. a dust storm
6. (any three) people out of work in the cities; industries shutting down; banks failing; food prices dropping; farmers losing their farms
7. They could not make a living on their farms, and they heard that life in California was easier.
8. The rain couldn't soak into the hard ground. So it ran off, creating gullies and washing off any remaining topsoil.
9. (any three) planting trees in windbreaks; planting grasses in gullies to stop erosion; subsidies paid for growing grass; dams and canals were built

Answers for the following will vary.

10. Too much land was plowed up to plant more wheat and take advantage of the high prices.
11. desperate; depressed; hopeless; tired of the constant dust and wind blowing; hungry

Additional Information on the Great Plains

- The story of the American Dust Bowl is told in John Steinbeck's book, *The Grapes of Wrath*. The book has also been made into a movie.
- Further land abuse occurred during World War II, as once again the pressure for more wheat was on. During this time, conservation priorities were largely set aside. Droughts in the 1950's brought conservation needs back into focus.
- Between 1934 and 1938, grasshoppers destroyed $315 million worth of crops. Poisons were used to kill them. People raked up huge piles of the insects from their yards. Dead grasshoppers made roads as slippery as ice, causing numerous accidents.
- Further droughts and dust storms hit the Great Plains in the 1950's and 1970's. Because of better management, the problems did not become as severe as they had in the 1930's.
- Droughts in the Great Plains appear about every 24 years.
- Water shortages in the western United States persist today. Much water used for farming is pumped from the Ogallala aquifer. Authorities fear this source may be pumped dry within a generation.

Drought

The American Dust Bowl

In the late 1800's, thousands of Americans moved west to the Great Plains. Some were in search of good land to farm. Many of these early pioneers settled on lush-looking lands in Nebraska, Oklahoma, Texas, and Kansas.

Before they could farm, they had to plow up the prairie grass. It was hard work. But the "sodbusters" kept at it. Soon, they had planted huge tracts of wheat and corn.

Other settlers became ranchers. They had plenty of prairie grass for their sheep and cattle. When the animals had eaten all the grass in one spot, the ranchers simply moved them to another spot.

The settlers didn't know it, but by destroying the grasses that held the soil in place they were setting the stage for a disaster that would come in the 1930's.

The Great Plains is an area that gets little rainfall. Usually it gets less than 15 inches a year. But the rainfall is not the same from year to year. Several dry years alternate with several wet years. In the 1890's and 1910's, periods of drought chased many settlers from the land.

Then came World War I (1914-1918). Farmers on the Great Plains were urged to grow as much wheat as possible to feed the troops. The price of wheat soared. Millions of acres of grassland were put to the plow. The weather cooperated by sending ample rainfall. Things were booming.

All this prosperity came to an abrupt halt in the 1930's. The entire decade brought severe droughts. Crops withered and died. There were no native grasses (and few crops) left to hold the soil in place. The Great Plains turned into the Great American Desert.

Winds began to sweep across the region. As they did, loose soil was picked up and carried for miles. Soon, all the fertile topsoil had been blown away. These terrible dust storms were known as black blizzards.

The dust storms caused enormous human suffering. The blowing dust blocked the sun at midday. People who had to go outside were forced to cover their faces to avoid choking. The dirt found its way inside houses, covering everything including people's food. Some people even died from breathing too much dust. The area suffered so many dust storms, it became known as the Dust Bowl. The dust blew across the country. It turned the clouds yellow as far east as New York.

The hot, dry weather created conditions that were just right for grasshoppers. Millions upon millions of them hatched. They destroyed the few crops that had managed to grow.

States affected by the Dust Bowl

(continued)

The American Dust Bowl *(continued)*

The problems of the Dust Bowl were compounded by the arrival of the Great Depression. During this time, millions of people in cities across the United States were out of work. Industries were shutting down. Banks were failing. Food prices were dropping to new lows.

Thousands of farmers across the Great Plains could no longer make a living on their farms. Many chose to migrate to California. They had heard things were better there.

Finally, the rain did return in the late 1930's. When it did, it caused more trouble. The ground was so hard, the rain could not soak in easily. It ran across the hard ground, creating huge gullies and washing away even more soil.

So, the federal government stepped in to help the Great Plains. The government gave loans to farmers who stayed on their land. It set up price supports, guaranteeing farmers a minimum price for their crops.

The U.S. Soil Conservation Service was set up in 1935. Its purpose was to teach farmers how to conserve the land. Trees were planted in rows called windbreaks, to block the force of strong plains winds. Grasses were planted in gullies to stop soil erosion (washing away of soil). Farmers were paid money called subsidies to let part of their land stay grassland. Dams and canals were built to bring water to the dry farmlands.

Today much of the damaged land has been restored. Droughts still occur on the Great Plains. But farmers have learned how to manage the land. So these droughts have not caused a disaster as severe as the Dust Bowl of the 1930's.

Name ______________________________

Date ______________________________

DROUGHT
REVIEW 1

Get the Facts

1. Why were early farmers on the Great Plains called sodbusters?

2. What were the two main causes of the destruction of the grasslands?

3. Why did World War I bring great prosperity for wheat farmers?

4. Why did the prosperity come to a halt in the 1930's?

5. What is a black blizzard?

6. Name three problems during the Great Depression.

7. Why did many Oklahomans leave their state for California?

8. What problems were caused when the rains returned in the late 1930's?

9. List three ways the damaged land was restored.

What Do You Think?

10. What part do you think greed played in the destruction of the grasslands during World War I?

11. Imagine that your family lived on a farm in Oklahoma during the Dust Bowl years. Describe how you think you and your parents might have felt during these desperate years.

Drought

Teacher's Guide for pp. 134–136

This teacher material covers the following reading selections:

What Is a Drought?
What Causes Drought?

Vocabulary (See page 144 for definitions.)

desert
desertification
drought
monsoon
semiarid land

Answer Key (Review 2, p. 136)

1. It can last for years and leads to massive deaths from starvation and disease.
2. Rainfall is far below normal over a length of time.
3. It must be far less than is normal for that area.
4. Deserts normally get very little rain.
5. During the good years, people can make a living there. When drought arrives, it brings much hardship.
6. the pressure of population
7. semiarid lands turning into deserts
8. (any three) overworking the soil; stripping land of trees and grasses; overgrazing; wasting water

Answers for the following will vary.

9. These people are very poor and would lose everything if they moved. They tend to stay in the area where they were born.
10. They could plan ahead by stockpiling food, digging wells, and building dams. They should conduct studies of the ecology of their country and implement farming methods which will conserve water and soil. They could aid people in building terraces, replanting forests, etc.

Additional Information on Weather, Geography, and Drought

- There is always the same amount of water on earth. But it moves from place to place in the water cycle. The heat of the sun causes water to evaporate from oceans and other bodies of water. This water forms clouds. Rain or snow falls from the clouds. It lands on the soil or in the oceans. The cycle repeats.
- The sun is the major influence on the earth's weather. It causes seasons, precipitation, and winds. Thus, changes in the sun's energy cycles bring changes in our weather, including droughts.
- Scientists can study the history of droughts by looking at tree rings. Thick rings mean good weather. Thin rings mean drought.
- Deserts and drought-prone areas are more likely to occur on the side of a mountain away from the sea. Clouds tend to drop their rain on the seaward side of the mountains before they pass over the mountains.
- Areas closer to the sea tend to get more rain. As clouds are carried inland for miles and miles, they gradually lose moisture. That is why many desert and arid regions occur at about 30 degrees latitude (see map on page 134).
- Drought is a problem in many developing countries. Many of these countries are located in areas where rainfall is not reliable. The lands bordering the Sahara Desert in Africa (for example, Ethiopia) suffer extreme droughts. China and India are two other countries that have lost millions of lives to droughts.

Drought

What Is a Drought?

Drought is not a sudden, spectacular natural disaster like a volcanic eruption or an earthquake. Drought creeps up slowly. It often goes unnoticed until water supplies dry up and crops fail. In fact, drought is the worst of all disasters because once it sets in, it can last for years. Prolonged drought often leads to mass starvation and death from disease. Throughout history, more people have died because of droughts and their effects than from any other kind of natural disaster.

Drought does not mean that an area gets no rain. It means that rainfall is far below normal over a great length of time. And the result is not enough water to support the life in that area.

The term drought means different things in different climates. If a desert normally gets less than 10 inches of rain per year, 10 inches is not a drought there. But in a rain forest that usually gets 100 inches of rain in a year, 20 inches would be a drought since plants there need more water to grow. Thus, droughts can happen anywhere.

When most people think of drought, they mistakenly think of deserts. Deserts are vast sandy areas with little or even no rainfall in a year. True deserts, which make up about one third of the world's land, have remained deserts for thousands of years. Few people live on a desert. So it is meaningless to speak of droughts in a desert.

Semiarid lands, which often border deserts, are actually the lands at greatest risk from droughts. These marginal areas get enough rain to allow some crops or livestock to be raised. In years of good rainfall, people can live fairly well on these lands. Populations grow, and more and more people move onto semiarid lands. But the drought years always come. And they bring great hardship to those living there.

Across the world, more semiarid lands appear to be turning into deserts. This process is called *desertification*. Some of this spreading of deserts may be due to changes in the climate of our earth. But people play a big part in this process. They overwork the soil. They strip land of trees and grasses. They allow livestock to overgraze. And they waste water.

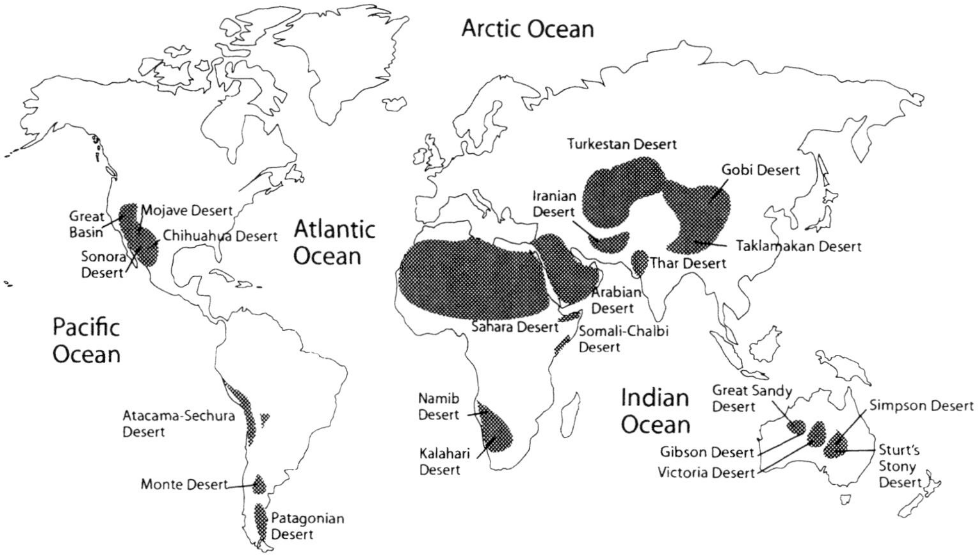

Earth's Deserts

What Causes Droughts?

Scientists don't yet understand all the causes of every drought. There are several different theories.

Many scientists believe that droughts are related to sunspots. When there is low sunspot activity, the weather is cooler than average. The oceans are not warmed as much as usual. With cooler oceans, less water evaporates into the air. With less moisture in the air, less rain falls on the land.

Other scientists say volcanic eruptions cause droughts. Huge amounts of ash in the air block sunlight and cause the oceans to cool. Some scientists believe that pollution has the same effect.

A third group of scientists point to the earth's magnetic field. Changes in this field alter wind and water currents. This causes rain to fall in different patterns.

Seasonal droughts are caused by the failure of the monsoons. In India and south Asia, the monsoon season falls between June and September. During this time, moisture-laden winds usually blow over the country from the Indian Ocean. Usually, the start of the season is very predictable. But when strong west winds blow over the Indian Ocean, the rains end up falling in the ocean. Farmers depend on the rainy season. When the rains do not come, their crops do not grow.

Droughts seem to occur in somewhat regular cycles. On the Great Plains, they arrive about every 24 years. In Africa, they appear at about the same interval. In China's Yellow River basin, they alternate with floods. Both kill a tremendous number of people. In the past century, 100 million people have died from droughts or floods in that area of China.

Meanwhile, the demand for water continues to rise. An ever-increasing population requires more water to live. Farming methods fail to make the best use of available water. Overplanting, overgrazing, and the cutting of too many trees sap the soil from what was once good farmland.

Name ______________________________

Date ______________________________

DROUGHT
REVIEW 2

Get the Facts

1. Why can drought be thought of as the worst of all natural disasters?

2. What does drought mean?

3. Why would 10 inches of rain in one year be considered a drought in one area but not in another?

4. Why is it meaningless to speak of drought in a desert?

5. Why are semiarid lands at greatest risk from droughts?

6. Why do so many people move into semiarid lands if there is risk from drought there?

7. What is *desertification*?

8. Name three ways people are causing desertification to occur.

What Do You Think?

9. Over the years, millions of people have died from droughts in India, Ethiopia, and the Yellow River Valley of China. Why do you think people don't just move out of these drought-prone areas?

10. In certain areas of the world, drought is expected every few years. How do you think the government of a drought-prone area could prepare its people for future droughts?

Teacher's Guide for pp. 139–141

This teacher material covers the following reading selections:

What Are the Effects of Drought?

How Can the Effects of Drought Be Controlled?

Vocabulary (See page 144 for definitions.)

aquifer	malnutrition
contour plowing	nomadic tribes
desalination	reservoir
fallow	terracing
famine	topsoil
groundwater	

Answer Key (Review 3, p. 141)

1. grasses, trees, and shrubs die; topsoil erosion
2. Crops fail and livestock die.
3. People weakened by malnutrition are more susceptible to disease. Also, people drink tainted water.
4. Modern transportation makes it possible to get food, medicine, etc., to remote areas.
5. store water for future needs
6. large source of groundwater
7. They create reservoirs, which can store large amounts of water to be used as needed.
8. (one) contour plowing; allowing land to lie fallow; using seeds that require less moisture

Answers for the following will vary.

9. (any two) desalination; NAWAPA; cloud seeding
10. Wealthy countries have the money to buy the food and water they need from other countries. Poor countries have little money and usually no stockpiles of food. They also have large populations, often in remote areas. The roads and transportation systems are poor, which hinder distribution of aid which may arrive. Poor countries have a large percentage of their population making a living from the land.
11. Stop lawn watering. Don't wash cars. Take showers rather than baths. Don't let water run while doing dishes.

Additional Information on Preventing Drought

- Historically, India, China, and Russia have suffered the world's greatest drought disasters. These countries get irregular rainfall and have had poor transportation and farming methods. Recently, however, they have made great strides in food storage and road building. These advances should substantially reduce drought-related deaths. Africa appears to be the problem area now. Ethiopia, Sudan, and Somalia have been devastated by droughts in the 1980's and 1990's.
- Huge dams have been built in the developing countries: a system of dams on the Indus River in Pakistan, the Aswan Dam on the Nile in Egypt, the Kainji Reservoir on the Niger in Nigeria, and the Volta River Dam in Ghana.

- Another way to prevent drought is by diverting water. For example, water has been diverted over the Rocky Mountains into the Great Plains area where it is needed. First, a dam and reservoir were built on the Colorado River. Then the water was piped over the mountains to the Platte River.
- Groundwater can be pumped from underground to irrigate fields. Much of the Great Plains uses this method today.
- Africa is currently one of the most drought-prone areas. One of the worst-hit countries is Ethiopia. It borders the Sahara desert, and it has one of the fastest growing populations in the world. A combination of drought, overpopulation, and civil war has caused great disaster there.

  Ethiopia was once a good farming area. But overpopulation has resulted in ruinous overuse of the land. Trees have been cut down to make more fields. Soils have not been allowed to rest in a fallow season. Because of this, crop yields have fallen to about one fifth of former yields.

  Ethiopia has suffered several major droughts in the last 20 years. Loss of life has been widespread. The country needs long-term help toward better farming techniques, improved transportation, and relocation of people to reduce future drought devastation.

What Are the Effects of Drought?

Effects on the Land

It takes years for the land to heal from a drought. Grasses and shrubs may return before long. But it takes years for forests to regenerate.

When grasses and crops die, nothing is left to hold the soil in place. Wind and water easily erode it. Topsoil is the upper layer of soil, which is used for growing crops. When this soil goes, it is not easily replaced. Scientists estimate that it takes 100 years to form one inch of topsoil.

Effects on People

When droughts occur, crops fail and livestock die. Soon, there is nothing for people to eat. Famine, or widespread hunger, is one of the most terrible effects of drought. If a country is wealthy, it may avoid disaster by importing food to feed its people. But most of the drought-prone countries in the world are also very poor. Without outside aid, these people are condemned to starvation.

Yet starvation itself is not the main killer during a drought. Most deaths are caused by disease. People weakened by malnutrition are more susceptible to disease. They also may drink water which has been contaminated. Then epidemics of disease break out.

Drought can also lead to more forest fires. During dry seasons, forest fires are more likely to occur and are harder to put out. The Great Chicago Fire of 1871 was more devastating because the town's wooden buildings were so dry due to drought.

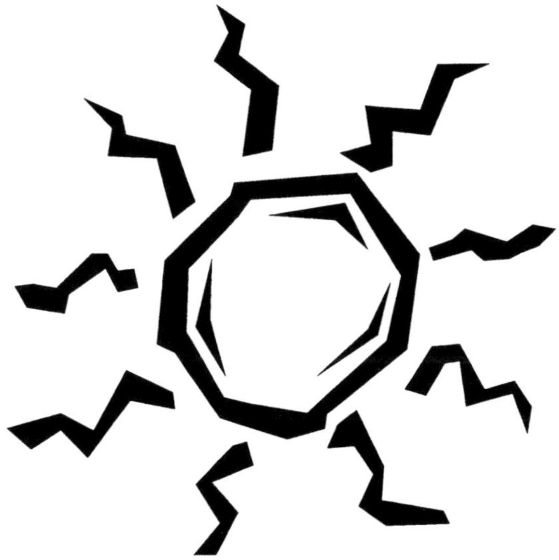

Drought goes on and on. It can last for years. Hunger and a feeling of hopelessness can be very hard to deal with. Food riots and lawlessness may break out.

When drought strikes a poor country, the government of that country usually has little money or food reserves. Throughout history, the people of these countries have died of starvation or disease. Today the outlook is somewhat better. Modern transportation makes it possible to get food, medicine, and other aid to countries in need. Individual countries and charitable groups organize relief efforts for those in need. Still, it is impossible to meet the needs of all these people in time. Massive suffering and death still occur. The suffering caused by drought is surely among the worst disasters on earth.

How Can the Effects of Droughts Be Controlled?

Storing Water for Future Needs

The best way to prepare for drought is to store water to be used in times of need. This is often done by building dams across rivers, creating reservoirs. In the United States, the Hoover Dam blocks the Colorado River to form Lake Mead in a dry area between Arizona and Nevada. In Egypt, the Nile River is blocked by the Aswan Dam, forming Lake Nasser. These lakes store vast amounts of water to be used as needed.

Another source of water is groundwater. Groundwater is water that has seeped down into spaces between underground rocks. Actually, more fresh water lies under the ground than in all the rivers and lakes in the world combined. Large groundwater sources are called aquifers.

Wells are used to get groundwater for drinking or irrigation. But there is a limit to the amount of groundwater that exists. In some places, people are pumping up groundwater faster than it can be replaced by nature. Some scientists fear that certain areas may be using up the groundwater.

People also save water for their own use by collecting rain as it falls. They collect it in rain barrels or stone jars.

Improving Farming Methods

Water can go further if farmers practice conservation methods. They can contour plow, rather than straight-line plow. That allows rain to sink into the ground better. They can plant cover crops to keep the soil in place, which prevents erosion. They can allow part of their land to lie unused each year. That means they do not plant crops. They allow the land to rest and build up moisture. They can build terraces on slopes to help rainwater soak into the ground instead of running off. They can also plant varieties of crops that require less moisture to grow.

Conservation of Existing Water

Individuals can conserve water during times of drought. They can stop watering lawns and washing cars. They can take quick showers instead of full baths. And they can install toilets that use less water per flush.

Developing New Sources of Water

In some areas, scientists seed clouds to produce rain. They drop silver iodide, dry ice, or various organic compounds into the clouds. The result is perhaps a 5 to 15 percent increase in rainfall.

The removal of salt from ocean water, or desalination, has been a popular idea for years. Unfortunately, the process is an expensive one. But it can be a good source of water. In Saudi Arabia, for example, desalted seawater is used for irrigation.

Plans for a massive water project were first proposed in 1964. The North American Water and Power Alliance (NAWAPA) would bring water from northern Canada and Alaska down to the United States and Mexico. The project would involve new reservoirs and canals leading south. It would be the largest water-diversion project in the history of the world.

Name ______________________________

Date ______________________________

DROUGHT
REVIEW 3

Get the Facts

1. Name two long-term effects drought can have on the land.

2. Why is famine the usual result of drought in poor countries?

3. Why do epidemics of disease often break out during droughts?

4. Why is the outlook in a drought better today than it was 100 years ago?

5. What is the best way to prepare for drought?

6. What is an aquifer?

7. How do dams help people in times of drought?

8. Name a farming technique that helps conserve water and soil.

9. Name two ways science may create new sources of fresh water.

What Do You Think?

10. Why do you think poor countries feel the effects of a drought so much more than wealthy countries do?

11. List all the ways you can think of to conserve water.

Additional Information

Major Droughts in History

Year	Location	Deaths
1337–1337	China	over 4 million
1769–1770	India	3–10 million
1837–1838	India	800,000
1856–1866	India	10 million
1876–1878	India/Madras province	3.5 million
1876–1879	China	9–13 million
1891–1892	Russia	450,000
1892–1824	China	1 million
1896–1897	India	5 million
1899–1900	India	1 million
1920–1921	China	500,000
1921–1922	Russia	5 million
1932–1933	Russia	5 million
1972–1974	Ethiopia	200,000
1973–1974	Ethiopia	400,000
1984–1985	Ethiopia	100,000

Drought

Activities for Students

1. Read John Steinbeck's *The Grapes of Wrath*. The book tells of the migration of a group of people to California during the Dust Bowl of the 1930's.
2. Find out about the United States Soil Conservation Service. What are its main purposes today?
3. Find out about the disasters that have taken place in Ethiopia, Sudan, and Somalia in recent years.
4. Read about deserts. How have plants, animals, and people adapted in order to live in these areas?
5. Read about the history of the Yellow River in China.
6. Read the story of the Aswan Dam in Egypt. Find out what problems have been caused by the construction of this dam.
7. Cloud seeding is a technique for causing rain and preventing hailstorms. Read more about this technique. Find out how it works and why it is controversial.
8. For years people have wondered about turning ocean water into fresh water. Read about the progress that has been made in this area.
9. Draw a picture to illustrate the water cycle.
10. Find out about the Cliff Dwellers, a group of Native Americans who lived in the southwest of the United States. Their civilization disappeared around the year 1300, probably due to severe drought conditions.
11. Read about one of the following droughts.
    - the 1876–79 drought in China
    - the 1769–70 drought in India
    - the 1865–66 drought in India
    - the 1921–22 drought in Russia
12. Read about how the United Nations works to provide aid in times of famine.
13. Some scientists think the climate of the earth is cooling. One effect of cooler weather is droughts, since less water evaporates from the ocean to form rain clouds. Find out more about this idea.
14. Overpopulation is a major factor in drought-related deaths. Feeding more and more people is difficult. One solution could be finding new sources of food from the sea, including seaweed. Find out more about this idea.

Drought Vocabulary

aquifer	large pool of groundwater that can be tapped with wells and pumps
black blizzard	dust storm that blocks the sun
contour plowing	plowing across slopes instead of up and down, in order to conserve water
desalination	process for removing the salt from seawater, making it fit to drink
desert	area that gets little or no rain
desertification	process of changing land into deserts
drought	period of time with much less than normal amount of rain
Dust Bowl	area of the Great Plains that suffered from drought in the 1930's
erosion	wearing away of soil, especially topsoil
fallow	when no crops are grown so land can rest
famine	widespread hunger
Great Depression	period in U.S. history when many were out of work, banks failed, etc. (1930's)
Great Plains	central plains states including Kansas, Nebraska, Texas, Oklahoma, etc.
groundwater	water stored beneath the earth's surface
gullies	channels in the ground caused by water running down a hill
malnutrition	poor health caused by poor nutrition
monsoon	rainy season of the tropics
nomadic tribes	tribes that moved from place to place
price supports	government payments that assure farmers of a minimum price for their crops
reservoir	lake formed by a dam
semiarid land	areas, usually found next to deserts, that receive unreliable amounts of rain
sodbusters	pioneer farmers on the Great Plains
subsidies	government payments to farmers that increase the money they get for crops
terracing	building walls to prevent rain running off land, usually on slopes
topsoil	upper layer of fertile soil, in which crops are planted
windbreak	row of trees planted to break the effects of the wind on the plains

POSTSCRIPT

Natural disasters are a basic part of the workings of our planet. As such, they will always be a part of our world.

Natural disasters take a tremendous toll, both in human lives and property damage. But there are benefits from their occurrence as well.

Earthquakes have taught us virtually everything that is known about the interior of the earth. Volcanoes produce fertile soil for crops. Thunderstorms and hurricanes bring rain. Floods spread fertile soil across the land. Forest fires cause new life to grow in the forests. Heavy snows melt in the spring and provide needed water for crops.

You may ask why, with our greater technology, we cannot control more natural disasters. Our ability to predict the occurrence of some disasters has improved greatly, allowing many lives to be saved. Disaster preparedness has improved dramatically. Yet we are not able to stop the progression of the natural disaster itself.

It may be that we are becoming more, not less, susceptible to natural disasters as time goes on. The world population is growing at a staggering rate. This means that more and more people are living in areas that are ripe for disaster. These areas include semiarid lands that are drought-prone, and places on or near earthquake faults.

No one knows where or when the next natural disaster might strike. A huge earthquake has been predicted for years along the San Andreas Fault in California. A strong hurricane scoring a direct hit on Miami or New Orleans could be devastating. Tokyo, Japan, could be hit by earthquake, tsunami, or volcanic eruptions. And the Bay of Bengal in India is likely to continue to be the scene of cyclones bringing terrible flooding.

In addition, human-caused disasters threaten us on a huge scale. Overpopulation, pollution of our air and water, destruction of the rain forest and other natural habitats, depletion of the ozone layer, and global warming could all be much more deadly to humans than the greatest natural disaster.

We will never be able to eliminate natural disasters. But we can lessen their impact by being aware of the dangers likely to occur in the area where we live and planning accordingly. We can also do what we can to reduce human-caused disasters by controlling our population, conserving natural resources, and in general living in tune with nature rather than thinking we are above it.

REFERENCES

Blizzards and Severe Winter Weather

Cable, Mary. *The Blizzard of '88*. New York: Atheneum, 1988.

Ebert, Charles H. *Disasters*. Dubuque, IA: Kendall/Hunt Publishing Company, 1988.

Robinson, Andrew. *Earth Shock*. London: Thames & Hudson, Ltd., 1993.

Droughts

Ebert, Charles H. *Disasters*. Dubuque, IA: Kendall/Hunt Publishing Company, 1988.

Knapp, Brian. *World Disasters: Drought*. Austin, TX: Steck-Vaughn Company, 1990.

Earthquakes

Archer, Jules. *Earthquake!* New York: Crestwood House, 1991.

Earthquake Safety Checklist. Federal Emergency Management Agency, 1985.

Safety Tips for Earthquakes. Federal Emergency Management Agency, 1983.

Fires

Ebert, Charles H. *Disasters*. Dubuque, IA: Kendall/Hunt Publishing Company, 1988.

Robinson, Andrew. *Earth Shock*. London: Thames and Hudson, Ltd., 1993.

Floods

Clark, Champ. *Flood*. Planet Earth Series. Alexandria, VA: Time-Life, 1982.

Ebert, Charles H. *Disasters*. Dubuque, Iowa: Kendall/Hunt Publishing Company, 1988.

Erlbach, Arlene. *Floods*. Chicago: Children's Press, 1994.

Frazier, Kendrick. *The Violent Face of Nature*. New York: William Morrow and Company, Inc., 1979.

Hart, Steven. "Waves of Terror." *Science World*, Vol. 51, No. 8, January 13, 1995, pp. 17–19.

Waters, John. *Flood!* New York: Crestwood House, 1991.

Hurricanes

Bixby, William. *Hurricanes.* New York: David McKay Co., Inc., 1979.

Erlbach, Arlene. *Hurricanes.* Chicago: Children's Press, 1993.

Harrist, Ron. "Hurricane Party Never Had a Chance." *Clarion-Ledger* (Jackson, MS), August 17, 1994.

"Hurricane." American Red Cross, 1994. (pamphlet)

Lehr, Paul E., R. Will Burnett, and Herbert S. Zim. *Weather.* New York: Golden Press, 1991.

Ludlum, David M. *The Great American Weather Book.* Boston: Houghton-Mifflin, 1982.

Tornadoes, Thunderstorms, and Lightning

Battan, Louis J. *Weather in Your Life.* San Francisco: W.H. Freeman & Co., 1983.

Erlbach, Arlene. *Tornadoes.* Chicago: Children's Press, 1994.

Frazier, Kendrick. *The Violent Face of Nature.* New York: William Morrow and Company, Inc. 1979. (pp. 46–76)

Robinson, Andrew. *Earthshock.* London: Thames & Hudson, Ltd., 1993. (pp. 117–152)

"Tornado." (pamphlet) National Oceanic and Atmospheric Administration, 1992.

"Tornadoes . . . Nature's Most Violent Storms." (pamphlet) National Oceanic and Atmospheric Administration, September 1992.

Volcanoes

Bender, Lionel. *The Story of the Earth: Volcano.* New York: Franklin Watts, 1988.

Johnson, Carl. *Fire on the Mountain.* San Francisco: Chronicle Books, 1994.

Kids Discover: Volcanoes. June/July, 1993. (magazine)

Knapp, Brian. *World Disasters! Volcano.* Austin, TX: Steck-Vaughn, 1989.

Palmer, Leonard and KOIN-TV Newsroom (Portland, OR). *Mount St. Helens: The Volcano Explodes.* Lee Enterprises, 1980.

Vrbova, Zuza. *Our Planet: Volcanoes and Earthquakes.* Mahwah, NJ: Troll Associates, 1990.

INDEX